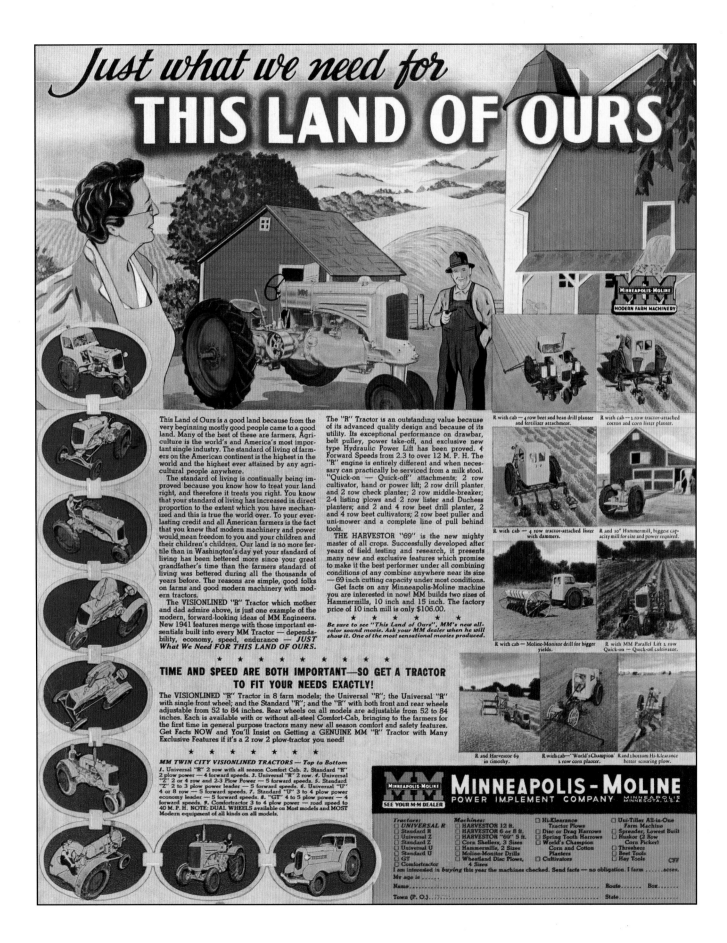

Motorbooks International

FARM TRACTOR COLOR HISTORY

FARM TRACTOR
Advertising in America
1900–1960

David Fetherston

First published in 1996 by Motorbooks International Publishers & Wholesalers, 729 Prospect Avenue, PO Box 1, Osceola, WI 54020-0001 USA

© David Fetherston, 1996

Motorbooks International is a certified trademark, registered with the United States Patent Office

The information in this book is true and complete to the best of our knowledge. All recommendations are made without any guarantee on the part of the author or Publisher, who also disclaim any liability incurred in connection with the use of this data or specific details

We recognize that some words, model names and designations, for example, mentioned herein are the property of the trademark holder. We use them for identification purposes only. This is not an official publication

Motorbooks International books are also available at discounts in bulk quantity for industrial or sales-promotional use. For details write to Special Sales Manager at the Publisher's address

Library of Congress Cataloging-in-Publication Data

Fetherston, David A.
    Farm tractor advertising in America 1900-1960 / David A. Fetherston.
        p. cm.
    Includes index.
    ISBN 0-7603-0162-X
    1. Farm tractors--United States--History. 2. Advertising, Magazine-United States.
    I. Title
TL233.6 F37F38   1996
629.225--dc20   96-21028

On the front cover: Ford Powermaster Series ad from 1958.

On the frontispiece: Willys Jeep ad from 1945.

On the title page: Minneapolis-Moline ad from 1941.

On the back cover: International Harvester ad from 1946 featuring the artwork of Donald Milles; Minneapolis-Moline ad from 1936; 1909 ad for the Reeves steam engine.

Printed in Hong Kong

Contents

ACKNOWLEDGMENTS

This book looks back across 100 years of tractors and the people who have been a contributing element to the American agricultural revolution. Its pages of history remind us that today's tractor and all its implements are due to the mastery of a small number of people with bright minds, creative dreams, and the financing to initiate them.

My love of machines goes back 40 years, to the first time I can remember rolling a toy car across the wooden floor of my room and to the six years of agricultural high school which introduced me to the many machines of farm life. From those high school years I thank Fred Fage and Jim Kilby, two of my agriculture lecturers who made learning fun while teaching me about mechanical agriculture, farming, and myself.

Thanks go to *Country Journal* (Cowles Magazines Inc., 4 High Ridge Park, Stamford, CT 06905) for the use of a quote on modern American farming from their September/October 1994 issue and to Hans Halberstadt for his help with some of the early ads.

Without the help of the editorial and production staff at Fetherston Publishing the idea, research, and execution of this book would never have produced such a bountiful crop!

For this work my thanks go to Gloria Fetherston for all her hours, at all hours, as advisor, manager, and editor on this project, to Nanette Simmons for her contributions as a writer and researcher, and to Cori Ewing for his research and writing.

David Fetherston

CHAPTER ONE

A BRIEF HISTORY OF THE AMERICAN TRACTOR INDUSTRY, 1900 TO 1960

Since the dawn of civilization, when hunters and gatherers found that nature's natural resources were no longer sufficient to feed their increasing numbers, man has turned to farming. Over thousands of years he has discovered seed collecting, planting, irrigation, and fertilizing, and in doing so, he also discovered that working the soil was never-ending and back-breaking. Necessity led him to find better ways to till the soil. Animals that were strong enough to pull plows were used in providing the power to till  a field or haul a load. But this practice has disappeared in the relative eye-blink of 150 years. The horse, the donkey, the farm laborer, and the slave all came before piston-powered farm machinery, as did its implements which made farming so much more productive.

Throughout the eighteenth century, tremendous changes took place in America. The first newspaper was published in 1704; France, Spain, and eventually Britain ceded their territories by the end of the Revolutionary War in 1783, and an organized and definitive government was set up with the Constitution of the United States in 1787. Families began to settle permanently in the newly formed States, and as the population grew, so did the need to research and develop ways to improve farming methods.

One invention in particular turned the course of farming history. The steam engine,

designed in 1780 by James Watt, a Scotsman, sparked the industrial revolution and introduced mechanical power to all ways of life, including the farm. Improvements brought about other inventions, and machines were developed that ran on a variety of fuels that ranged from wood-fired steam engines to kerosene, gasoline, diesel, alcohol, and eventually to LP gas.

By 1800 the British had developed a machine to harvest grain, but it was not entirely effective. New ideas for mechanized farming were not a great hit with farm laborers as they felt jobs would go to machines instead of to men; however, the future of farming was mechanization.

In America, Cyrus Hall McCormick had read everything he could about harvesting machines and had worked alongside his father who had tried for many years to develop a harvester. Young Cyrus took this collection of concepts and made them work. In 1831 he demonstrated the first practical working horse-drawn reaper on a farm in Walnut Grove, Virginia, and three years later received a patent for his work.

The financial panic of 1837 nearly wiped him out, but through some good manage-

ment and some of the first farm machinery advertising, he continued his "Virginia Reaper" business, selling 50 machines a year.

Cyrus moved west to Chicago and his business grew steadily by selling an improved version of his reaper. He could see a potential vast sea of grain flowing from the millions of acres of unimproved land in the Midwest. By this time he had refined, strengthened, and improved his reaper to the point where it was a machine that could be used all day without any problems—his Virginia Reaper was capable of cutting 20 acres of wheat a day.

When he took a version to England for demonstration at the great Crystal Palace Exposition, the British press recognized that his reaper was a machine of enormous value to agriculture, and Cyrus was awarded a gold medal. His advertising and favorable press reports made the selling of his reapers simple.

Back on the farm, McCormick was among the first to offer farmers direct credit for buying his machinery. Through this financial revolution, American farmers got a chance to improve the quality of their lives and create larger, more profitable, farms. Other individuals and companies parallel this story, but none made as large an impact on the business of American farming in the late 1800s.

There was another revolution still to come. Use of the external combustion (steam) engine in industry and shipping was just beginning. On the farm, the steam engine began life as a stationary unit pumping water or powering machinery, such as threshers. Soon people realized that it would be more efficient "to take the engine to the job." Mobility was added and the "traction engine" was born.

However, farmers were not satisfied and insisted that the engines be able to pull loads and draw plows. The result was steam-powered behemoths that could plow vast fields. Despite the new machines' speed and relative efficiency, Clydesdales and other beasts of burden still had job security. The tractors were expensive and required a team of men to operate them, precluding their use on any small farm.

N. A. Otto is generally recognized as the inventor of the internal combustion engine. In 1876 he developed the first practical unit and made a number of improvements in the following years. Otto's patents excluded its use by other companies until the patents' expiration in 1890.

By this time many small newspapers and magazines dedicated to farming had appeared. *The Pacific Rural Press* showed up in California in 1870 and it was from this type of publication that farmers, especially in California, came to know about the business of farming and what implements and machines were available. Numerous early stationary engine and traction engine manufacturers in the West advertised in this newspaper. It is still being published, over 120 years later, as the *California Farmer.*

Creativity was plentiful. In 1892, John Froelich mounted a Van Duzen gas engine on a re-engineered Robinson traction engine. It is most likely that this was the first liquid fuel, engine tractor that worked efficiently. These new gas tractors proved to be a boon for farming. Instead of running on coal or wood, the new tractors used a portable liquid fuel, which not only produced more power per cubic measure, but was cleaner and more powerful. In an amazing show of confidence, Froelich had his

machine pull a thresher for 50 days in a wide range of operating temperatures to prove its ability and reliability.

The following year Froelich started the Waterloo Gasoline Traction Engine Company and put a landmark on the history of farming in America. Others who saw its potential started delving into this new liquid-fuel power including the McCormick Company, which built its first gas engine in 1897.

In 1898 Huber Manufacturing switched from producing steam engines to gas engines by buying the rights and interests of the Van Duzen Company which had been producing a version of John Froelich's new tractor. With this technology available the Massey Harris Company built the four-cylinder "Wallis Bear" in 1902. It was a giant of a machine that could do the work of 40 horses by pulling 10 14-inch plows.

The International Harvester Company was formed in August 1902 from an amalgamation of McCormick and Deering, both famous for their durable grain harvesters. By 1924, the International Harvester Company had produced its first efficient all-round tractor, the Farmall.

The Gray Tractor Company, owned by Chandler Knapp, produced its first two-cylinder model in 1908 as an alternative to horse-drawn machinery in orchards. By 1914 Knapp had relocated the company to Minnesota and continued production there until 1933.

In 1909 another notable tractor emerged with the first Rumely OilPull, which was designed by John Secor in a deal with Edward Rumely. For years, Secor had been working on the development of an engine to burn low-grade fuels. OilPulls were extremely successful nationwide with 14 different models marketed until 1931.

Advance-Rumely took to advertising like ducks to water. They advertised their tractors extensively in newspapers such as the *Iowa*

Threshermen, The Farmer, The Iowa Homestead, Nebraska Farm Journal, and *The Iowa and Corn-belt Farmer,* as well as in *Country Gentleman, California Cultivator* (formerly *Pacific Rural Press*), and *Wallaces' Farmer.*

These newspapers and magazines had an enormous impact on the sales of tractors. Not only did they advise farmers of what was new to the business, they offered ways to finance equipment, carrying extensive advertising programs from all the tractor and implement manufacturers. Advance-Rumely, Avery, Deere, Samson, Case, and Twin City all carried out broad advertising campaigns in these publications.

One of the great advances in tractor engineering came out of Stockton, California. It was here that the track tread was invented just before the turn of the century by Benjamin Holt. His company, Holt Manufacturing, had been in the tractor business since 1890, originally as the Stockton Wheel Company building steam-powered tractors.

Holt's bright idea was to invent a new drive system that used a flexible metal track in lieu of round wheels. These tracks spread the weight over a far greater area and lightened the compression load on the soil. This allowed the machine to traverse much softer soils and steeper grades than would otherwise have been possible.

In 1909 variations and novelties in tractor design were plentiful, and the Avery Company, previously a steam tractor producer, introduced its unique Tractor Truck. This truck could carry a load of three tons or could haul three 14-inch plows and the wheels could be changed for the varying surfaces on which the machine operated.

By 1910, a total of 28 million combined animal and mechanical horsepower was used on American farms—but only a quarter was mechanical horsepower, developed almost entirely from steam. Gasoline tractors at this time were generally four-stroke, had automatic valves, and dry batteries for providing ignition during startup. Spark was provided by a magneto or generator once the motor was running. The frames were built of iron channels and the transmission gears were of cast-iron and were generally exposed.

In 1915 the Moline Plow Company bought the Universal Tractor Company and began producing the Moline Universal with a four-cylinder engine, electric starter, and optional lighting. It was one of the first cultivating tractors, a strange-looking device much like a giant garden cultivator with two huge drive wheels up front.

Tractors worked well for the initial groundbreaking on large farms but would not work for everyday cultivation as they couldn't stay within the rows. Horses were still needed and required for some work on most moderate-sized farms.

Until this time most tractors were still too large and cumbersome for one man and would make a mess of any planted field. The rapid decrease in the size of tractors allowed farmers to see the possibility that tractors, not horses, would become the power source of the future farm. Smaller row-crop farmers still needed draft animals. However, many small companies, especially in California, were building machines suited to a one-man operation for row-crop and orchard work.

Responding to the demand caused by World War I and the new technologies and designs available, hundreds of companies began to produce machines that were adequate for a small farmer's needs and budgets. But World War I took men and horses away from farms, and tractors were soon needed to supplement the remaining human and animal labor force.

The military leadership of World War I didn't consider trucks reliable and decided to use only horses to move their implements of war. The mortality rate for war-time horses was high and the farmers returning from war realized that they needed replacements for their lost horse-power.

General Motors entered the tractor market at this time with the purchase of the Samson Sieve Grip Tractor Company in

1914. The company began construction of a factory the following year and evolved its own Samson tractor within a short time. It was an evolutionary design that was most useful on the tule and peat soils of California reclamation areas.

Henry Ford's new small Fordson tractor took the farming world by storm. Ford had previously built an Automobile Plow in 1907, which used parts from the Model B and K Ford cars, but it did not live up to expectations and was soon abandoned.

The 1917 Fordson was built as a unique tractor design, but it had a four-cylinder engine that set new records for difficulty in starting. A high-maintenance ignition plus a carburetor that delivered only a little fuel gave it a cantankerous start-up, although starting the engine on gasoline before switching to kerosene helped. Three forward gears allowed speeds up to 6 mph. Electric starting and lighting were options. The worm gear drive was infamous for heating the tractor's metal seat to a level of extreme discomfort. Ford claimed that the Fordson was the first tractor to use a unit frame, even though the Wallis Cub was really the first. Most companies eventually adopted the unit frame design.

Toward the end of the 1910s there were a numerous corporate buyouts. Large tractor companies bought out smaller companies to obtain design patents, sales distribution, and manufacturing facilities in order to increase their sales potential and corporate value. This revolution in the tractor business led to a flood of new tractor designs and stock sales to fund their operations.

Unfortunately, some operators had no intention of building workable tractors and were just out for a quick buck, selling inferior designs. The Ford Tractor Company from Minneapolis was most likely one of these. The company had absolutely no connection with Henry Ford—rather, the company was trying to take advantage of the fact that farmers were

expecting a good inexpensive tractor from Henry Ford and used the name "Ford" for a tractor before Henry Ford did! The company had the right to use the name "Minneapolis Ford Model B" and even went so far as to hire a man named Ford.

Many farmers were scammed by purchasing poorly designed tractors, but this ceased when a Nebraska legislator, Wilmot F. Crozier, bought a Minneapolis Ford Model B that broke down as soon as he got it to his farm. He then bought a used Rumely OilPull that performed well. Realizing he'd been a victim of fraudulent advertising, Crozier decided to put charlatan dealers out of business. In 1919, with the help of L. W. Chase (the former head of agricultural engineering at the University of Nebraska), Crozier authored a bill that required any company wishing to sell tractors in Nebraska to pass an operational test. This tested the tractor's horsepower rating, oil, water, and fuel consumption as well as its mechanical soundness.

When the test program became law, the first tractor, a Waterloo Boy Model 12-25, was tested in April 1920. This legislation was one of the earliest forms of consumer protection and proved very effective. Successful passage through this testing certified the rating of a company's tractor and this point was well taken up in advertising for many years.

There was an immediate effect on all tractor horsepower rating numbers with this legislation. Horsepower ratings were always given in pairs such as the Avery 40-80. The first number was the pulling power on the draw bar (used for attaching plows), with the power at the pulley (used for operating separate farm machinery, such as threshers) the second number. The test results had to match the company's figure claims, and if they did not, the tractor's rating was changed to reflect the actual power. The Avery 40-80, for example, was changed to the Avery 45-65, reflecting not only over-rated drawbar power but underrated pulley power.

This mandatory testing significantly cleaned up the tractor business,

allowing legitimate companies to market reliable and, eventually, inexpensive tractors.

It was Henry Ford who took the trend toward affordability of a tractor a step further. With Ford's mass production ability, the Fordson tractors were inexpensive enough to force a number of competitors out of business. Those companies that managed to stay in business had to readjust their prices to new competitive lows, a task that unfortunately eliminated the profit for numerous sellers. But the price wars were a great boon to farmers, many of whom could now afford to purchase tractors.

During the early to mid-1920s, sales increased steadily and a number of tractor companies felt secure in spending time and money refining their products, while tough competition forced them to present the best product possible. A general trend toward improved engineering and higher quality was seen at this time.

Air cleaners became common accessories in the 1920s as people began to realize there was enough gain in performance to offset any restriction that a dust filter incurred. A test of 26 air cleaners by the College of Agriculture in Davis, California, demonstrated the effectiveness of the air cleaners and led to improved designs.

International Harvester continued to be a major force in the tractor industry. Introduced in 1924, the company's new Farmall tractor was phenomenally successful—it is a popular tractor even to this day. This model is considered the first modern all-purpose "tricycle" farm tractor.

The C. L. Best Tractor Company and the Holt Manufacturing Company merged in 1925 to form the Caterpillar Tractor Company. The year 1926 was a great one for the tractor business with over 50,000 American tractors exported around the world.

Tractors, many now used as all-purpose machines, were further adapted for cultivating and some, like the John Deere GP, had arched front axles to increase clearance and allow the straddling of rows. The GP also had the industry's first power lift for raising drawn implements off the ground.

A moment's domination of the market doesn't guarantee success, and in 1928 Ford stepped out of the tractor business until the late 1930s as the International Harvester Company was able to beat Ford at his own pricing game.

In 1871, Thompson's Rubber Tired Steamer introduced the rubber tire using rubber cleats on the wheels. This design did not meet with great success and it wasn't until 1928 that rubber tires were used successfully. In Florida farmers were having problems with the steel wheels of their tractors damaging the roots of their orange trees. As a remedy they began to fit 8x40 truck tire casings to their wheels.

In 1931, B. F. Goodrich Company brought out a "zero pressure" tire made of rubber arches over a steel base that allowed fitting the tires to steel wheels. The company claimed that the flexible arch provided better contact than either a steel or a pneumatic tire could. One benefit that was indisputable was that the nonpneumatic tire was not subject to puncture.

Firestone was one of a number of companies that began experimenting with rubber tractor tires in 1932. By 1940, 95 percent of all new tractors were fitted with pneumatic tires, and shortly after the end of World War II virtually all tractors left the factory fitted with rubber tires.

The advantages of rubber tires were astounding. They provided higher fuel economy, increased speed and equal or better traction, depending upon conditions. The speeds possible on rubber-tired tractors were so much higher that Barney Oldfield, the famed Indy 500 driver, received a speeding ticket on one during the 1930s!

Metal weights had been commonly hung on wheels prior to the introduction of rubber tires to provide more traction. Problems began with the increase in speeds and the lighter weight of the rubber tires. Water-filled tires solved the problems and replaced the metal weights. It was also found that the water provided shock absorption as well as traction. To prevent the water from freezing, calcium chloride was added as an antifreeze agent.

Since the late 1910s, there was a slow trend toward a general purpose tractor. Tractor production reached an all-time peak in 1929 with 229,000 units. The number of crawlers, or track-laying tractors like the Caterpillar, increased drastically in the second half of the 1920s. Because of the use of crawlers in mining, road construction, industry, and agriculture, their production was not hit as hard when the economy took a sharp turn. With the onset of the Great Depression, American tractor production fell from more than 202,000 units produced in 1930 to less than 72,000 in 1931, and the following year tractor production dropped to an all-time low of 19,000 units.

Yet despite the Depression and pitiful tractor sales, development was still taking place. Rubber tires and diesel engines were the most important. Caterpillar had one of the first diesels in 1931, and by mid-decade a number of other companies had introduced diesel models. Adjustable track width was another innovation available from numerous companies.

The late 1930s introduced sleek styling to tractors and words like "streamlined" were borrowed from automotive advertising. More comfortable cabs were introduced. At the 1938 tractor show, the Minneapolis-Moline Company introduced two UDLX models. Also known as the "Comfortractor," they were touted to be the equivalent of an automotive sports coupe and a sports roadster.

Minneapolis-Moline was among the first to start advertising in full color. Their mid-1930s ads were very colorful and brimming with details. This type of advertising became the basic element for virtually all Minneapolis-Moline's advertising in the years to come.

Ford returned to the tractor market in 1939 with the Ford-Ferguson. This new tractor had more power, was easier to use, and had a unique hydraulic plow-lift with a three-point hitch design that enabled the use of a larger plow on the smaller tractor.

Henry Ford and Harry Ferguson had a verbal agreement to build and sell this tractor together, but Henry Ford, anything but a gentlemen where profit was involved, reneged on paying Ferguson royalties, forcing a large legal battle. Ferguson was no angel either and the suit dragged on from suit to countersuit. Ferguson eventually won a victory in 1951 by accepting a settlement of $9.25 million, although Ferguson felt it was actually a defeat as the settlement was only a small percentage of his original demand.

World War II suspended tractor experimentation and brought about a shortage of materials. Steel replaced bronze and copper in some applications, making some machines inferior to their prewar equivalents. As a result of this and the conversion of manufacturing plants to war use, preventative maintenance was stressed on the farm to keep old equipment running. Old tractors were recycled for the war, creating as a result, a shortage of some old tractor examples today.

After World War II, the demand for tractors was much higher than the producers could possibly meet. Postwar farming returned a far higher income and European farmers, receiving money under the Marshall Plan, wanted to spend it on tractors.

Advertising played a significant role in the sales arena. Ford, Oliver, Case, and International Harvester had a variety of advertising styles in those postwar years. Magazines such as *Successful Farming*, *Farm Journal*, and *California Farmer* carried ads from these companies that used a lot of illustrative work rather than hard-edged photos of tractors at work. This created an ethereal image "of man, doing man's work" as the great American farmer on the perfect piece of land.

Oliver, in particular, created this mystical imagery in its advertising using only two colors. However, as the paper and print quality offered by the magazines rose rapidly in the early 1950s, International Harvester, Allis-Chalmers, and Case soon began using full color for all their print advertising.

Four-wheel drive, common in today's tractors, was tried out in the early part of the century, but due to weight and complex problems, it was not until the engineering of the 1960s that it became a reality along with mechanical marvels such as the turbocharging and air conditioning on modern tractors.

Many tractor companies were in business, but only a few of the original innovators are still around today. It was unimaginable in the first half of the century that the second half would see foreign companies becoming players, if not sales leaders in the American tractor market.

By the end of the 1970s Deere & Company had replaced International Harvester as the overall tractor sales leader, and Case was part of Tenneco, a Texas oil conglomerate. The late 1970s were anything but kind to International Harvester. A combination of competition, diminishing tractor sales, and a UAW

strike led to multibillion dollar losses and its eventual sale to Tenneco, who merged it into a new division called Case-IH.

With improved shipping and efficiency, most farmers have improved their level of crop specialization, and this in turn has led to individualized equipment, including a mass of planters and pickers for crops as diverse as lettuce, cotton, and pineapples. Also, the average farm size had been increasing, and farmers began to require larger equipment.

According to the September/October 1994 *Country Journal*, "Farmland today is responsible for feeding, clothing and housing 250 million people in the United States and millions more abroad. It is the foundation for our food and fiber industry, which provides jobs for 20 percent of the work force and contributes $820 billion to the gross national product. Farmland also provides open countryside, offers food and habitat for diverse wildlife and maintains a link with the nation's past."

The tractor has had so many peaks that it can not be said that its heyday has arrived yet. Tractors are better than ever today. They are reliable, fuel efficient, and power and traction have increased tenfold. The modern farm tractor has been one of man's greatest mechanical achievements. It has allowed him to turn unusable land into productive acreage while at the same time dramatically increasing the productivity of arable land.

The tractor's complexity has increased tremendously. Today, its amazing productivity equals its reliability as a tool for the good of humanity, unrivaled in helping the world feed its people.

CHAPTER TWO

ALLIS-CHALMERS

The Allis-Chalmers Company was formed in 1901 following the merger of four companies: E. P. Allis & Company (a flour milling supplier) of Milwaukee, Fraser & Chalmers of Chicago, Gates Iron Works of Chicago, and the Dickson Manufacturing Company of Scranton and Wilkes-Barre, Pennsylvania.

Almost 20 years before this merger, Edward P. Allis had worked on improving the steam engine as a power source; in 1884 he built the largest centrifugal pump in America. Two years later he produced the Allis Triple-Expansion steam engine. The purchase of Bullock Electri-

cal Manufacturing of Cincinnati, Ohio, expanded the company's horizons into the electrical machinery business, and in 1906 Allis-Chalmers had its first steam turbine-generator.

The next major product line introduced was the tractor in 1914 which blossomed into a separate Tractor Division 12 years later. Allis-Chalmers' first machines were three-wheeled. The initial model was the 10-18, designed with a traditional tricycle layout. It was quickly followed by the Allis-Chalmers version of the Moline Universal tractor which had two large drive wheels up front with the driver seated

at the rear of the machine, over the implement. This version continued in the line-up until 1923.

The Allis-Chalmers 18-30 model, which appeared in 1919, barely survived the post–World War I depression and the stiff competition from Fordson and International Harvester in the tractor-saturated market of the 1920s. However, it was of conventional four-wheeled design and sold steadily until it was re-engineered and reintroduced as the Model L in 1929.

During the financial confusion at the end of the 1920s, farm equipment companies closed their doors at a rapid rate. Many companies

This December 1920, ad for the Allis-Chalmers 18-30 uses great line art. The headline "The Brute Power of 15 Horses" sounds quite funny today. This model ran in production for 10 years and sold approximately 16,000 units.

In 1921 Allis-Chalmers added more detailed line art to their advertisements noting, "How a $42,500,000 Company Upheld a 65-Year Reputation." Their claim was backed with a small note and illustration that each engine underwent a severe 10-hour block test on a dyno.

disappeared, but other distressed companies were picked up by financially able competitors for their debts or for pennies on the dollar. In this way, Allis-Chalmers added Monarch Tractor Corporation in 1928 and the La Crosse Plow Company.

Allis-Chalmers, with the purchase of the Advance-Rumely Thresher Company in 1931, continued to expand its manufacturing capabilities to include tillage implements and harvesters. This purchase made Allis-Chalmers the fourth largest farm equipment maker in the United States. In 1938 they also added Brenneis Manufacturing from Oxnard, California, which came with a line of tillage implements for row crop farming.

The West Coast connection prompted a new color image for the company products—the wild poppies on the hillsides were so impressive that Allis-Chalmers changed the color of its tractors from a somber green to Persian Orange.

Tire manufacturer Harvey Firestone was determined to improve the tractors' traction using adjustable air pressure and different tread patterns. His tires were also responsible for a great reduction of vibration, which in turn increased the comfort of the tractor operator. Allis-Chalmers was impressed with Firestone's invention and released the WC row crop tractor as the first tractor with inflatable rubber tires as standard equipment.

Technological innovations like the inflatable tires helped Allis-Chalmers leap forward in farm tractor sales. Very quickly, every other manufacturer moved to offer these tires, first as optional and then as standard equipment.

Allis-Chalmers continued to expand and introduced the Model C in 1940. By 1948, 70,000 of these tractors had reached the market. The 3 3/8x3 1/2-inch engine developed 18.43 drawbar and 23.30 brake horsepower with a maximum drawbar pull of 2,206 pounds.

The model line-up evolved and expanded again in the early 1950s with the WD-45, an improved version of the WC. The 1950 version developed a maximum drawbar pull of 4,304 pounds. A-C advertising boasted five engineering advancements: the Power Crater engine, the automatic traction booster, power-shift wheels, a two-clutch power control, and a snap coupler for implement attachment. From 1953 to 1957, over 83,000 WD-45 tractors were produced in both gasoline and diesel models.

The desire for a larger share of the market once again had Allis-Chalmers out shopping, and in 1955, acquired Gleaner, the Independence, Missouri, harvester maker. Ten years later Simplicity Manufacturing was added to the group. By this time the company owned 26 facilities and manufactured a diverse line-up of farm machinery.

Allis-Chalmers experienced 30 years of dynamic leadership from 1955 until 1985, when the agricultural equipment market began to plummet. The decision was made to sell out to a subsidiary of Klockner-Humboldt-Deutz ERG of West Germany. However, 5 years later American investors were able to re-acquire the original Allis-Chalmers Agricultural Group.

This 1923 ad for what appears to be excess stock of 10-18 horsepower tractors was part of a larger ad for a farm machinery distributor. The 10-18 was powered by a two-cylinder horizontal engine and rode on a tricycle framed chassis.

One of the first Allis-Chalmers crawler type tractors was the Model M. The ad featured a screened photograph of the crawler with a disc plow attached and noted factory branches in Oakland, Los Angeles, Portland, Spokane, Pocatello, and Billings.

This 1937 two-color ad for the Allis-Chalmers WC was bannered with "Buy The Leader." The ad noted that the WC was offered with the air tire option for $925.00. Soon after the price dropped $100 for this version. It was the first general purpose tractor mounted on rubber to be officially tested in the Nebraska tractor tests.

BETTER LIVING
follows <u>tractor</u> power

Pictured here are symbols of two generations on the farm — the old hand husking peg and the modern tractor-powered corn harvester.

To most farmers, the husking peg brings vivid memories of aching muscles and hands cut by the sharp-edged husks. Slash ... twist ... and a fling sent the ear flying to the wagon. Those motions, 10,000 times repeated, were necessary to pick the ears from a single acre.

But all this has changed in a single generation. Today, the Allis-Chalmers Corn Harvester, shown here, delivers ears to the wagon faster than ten men husking by hand.

Tractor-powered equipment has reduced the work time necessary for growing corn to less than five minutes per bushel. Corn is not only processed into many important foods but is a basic feed crop for our great livestock, dairy and poultry industries.

Tractor power is the symbol of abundance and better living — in the country and in the city.

Farm and Industrial Wheel Tractors • Farm Equipment •
2-Cycle Diesel Crawler Tractors • Road Machinery • Engines

ALLIS·CHALMERS
TRACTOR DIVISION • MILWAUKEE 1, U.S.A.

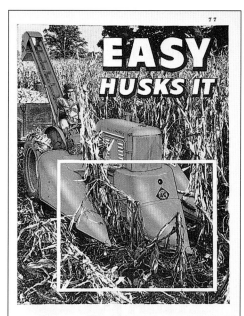

EASY
HUSKS IT

Here is picture evidence of what it means to have your own Corn Harvester, rolling out the corn during those few days of favorable husking weather.

A week later and the deterioration of this corn could have been severe, waiting on a custom machine. Borer-weakened stalks within the white square are already brittle and breaking over.

But watch the gentle skill of the Corn Harvester. See those long, low-sloping gathering snouts? Skimming the ground, they'll glide under that low-hanging ear. Spider wheels with curved fingers will gather it in. Rubber husking rolls and spring steel "husking pegs" will strip off the husks gently . . . like the human hand.

ONE MAN *alone* can attach the Corn Harvester to the Allis-Chalmers WC tractor in less than 30 minutes.

See it operating and you'll know that here is a new advanced method of *easier* corn harvesting for the family farm.

ALLIS·CHALMERS
TRACTOR DIVISION — MILWAUKEE 1, U.S.A.

The Allis-Chalmers corn husker fitted around the front of models including the WD-45 and could harvest two rows of corn at a time, dumping husked corn into a trailer pulled behind the tractor.

By 1949 the line-up of implements expanded at Allis-Chalmers to include industrial tractors, diesel crawlers, road machinery, stationery engines, and corn harvesters, like this one that "could work faster than ten men working by hand."

The Dynamic Power-Crater Engine powered the WD-45 tractor, running on regular gasoline. It also featured a traction booster, power shift wheels, and a snap coupler. Between 1948 and 1953, 131,000 units were built of this obviously popular tractor. During the next three years another 83,000 were built. This color ad from 1953 emphasized "the explosive power of the Power-Crater Engine."

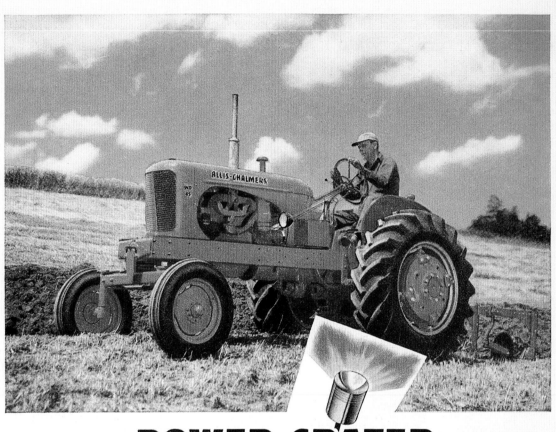

DYNAMIC POWER-CRATER ENGINE

powers the great new WD-45 tractor

Listen . . . its sound tells you something big has happened in tractor engineering!

Watch three plow bottoms bite into your toughest soil. The hydraulic Traction Booster bears down with added weight on the rear wheels. The new POWER CRATER engine pours on extra horsepower. THEN—the new WD-45 tractor really talks!

Farm faster . . . deeper! Handle heavier loads, with rugged new helical gear, 4-speed transmission. Get that extra power you have always wanted, *yet save several hundred dollars on the price of your tractor.*

Mark that name, POWER CRATER. It's power to prosper. It's *yours* . . . in the dynamic new WD-45!

Ask your dealer to
DEMONSTRATE *these*
5 Great Allis-Chalmers
Engineering
Advancements

POWER CRATER ENGINE
introduces high-compression turbulence, Center-Fire ignition, high-octane performance with *regular* gasoline.

AUTOMATIC TRACTION BOOSTER
increases weight on drive wheels to meet the load.

POWER-SHIFT WHEELS
use engine power to space rear wheels quickly and easily.

TWO-CLUTCH POWER CONTROL
stops or slows forward travel to ease through tough loads; lets power-driven machines continue running.

ALLIS-CHALMERS
TRACTOR DIVISION · MILWAUKEE 1, U. S. A.

Plus—SNAP COUPLER! *Handiest implement attachment ever devised. Ask your dealer!*

Announcing

the great new

SIX FOOT

ALL-CROP Harvester

This month the curtain lifts on a new, greater capacity harvest for the home farm.

It's a new ALL-CROP Harvester with many advancements that bring more pleasure and profit to harvest time.

The new Model 66 will not seem a stranger. In it you will find all the desired features that have made the ALL-CROP Harvester known and admired throughout the world.

In addition, the Six-Foot ALL-CROP Harvester has many new abilities. It will handle two wide rows of soybeans or sorghums. Crops flow in a broader, WIDE FLOW stream to the famous rubber-cushioned bar cylinder.

A New STEP-UP straw rack boosts capacity, gives straw a faster, rougher ride; separates cleaner.

Out of the experience of many thousands of owners over the years come these new features that make this the greatest ALL-CROP Harvester of all. See it soon at your A-C dealer's — get the latest story on harvesting over one hundred cash and conservation crops.

Hear farm news — music — markets
NATIONAL FARM & HOME HOUR — NBC — every Saturday

Features for the Harvest You Have Always Wanted

New SIX-FOOT header.

New sturdier SIX-BAT reel, for smoother feeding action.

New STEP-UP straw rack, handles a heavier volume, separates faster and cleaner.

New rotary flail-type Straw Spreader — optional extra equipment.

New Center Suspension Spring for header mounting, with quick-cleanout openings on lower draper housing.

Hydraulic header lift, controlled from the tractor seat.

ALL-CROP is an Allis-Chalmers trademark

ALLIS-CHALMERS
TRACTOR DIVISION • MILWAUKEE 1, U. S. A.

pay dirt farming

TO BETTER LIVING
TO BETTER FARMING
TO MORE PROFIT

WITH ALLIS-CHALMERS HYDRAULIC-ACTION B AND C TRACTORS

The wisest idea of 1950 may be for you to make your place a two-tractor farm.

Here is a choice of two tractors that will fit your budget. Either costs less to own and operate than horse equipment. Both have the last word in hydraulic finger-tip control . . . with Quick-Hitch front-mounted implements priced lower than horse tools.

Most economical cultivating tractor ever built, the 2-row Allis-Chalmers Model C lets you *see*. Relax on the wide cushion seat and watch the rows ahead, with Dual Depth Control gauging each gang depth accurately and independently.

The same aggressive, versatile power is yours at even lower cost in the standard type Model B, shown below with Full-Vision V-Belt Mower. A wide line of full-vision C or B implements is available, all with hydraulic control.

Your family farm can hit pay dirt, with two tractors doing two field or chore jobs simultaneously. It's an idea to talk over with your Allis-Chalmers dealer.

ALLIS-CHALMERS
TRACTOR DIVISION • MILWAUKEE 1, U.S.A.

Quiet V-Belt Mower has a new vari-speed drive with high strength V-Belt which absorbs shock, vibration and noise. Front-mounted in full view on either the B or C Tractors.

In 1950 this Allis-Chalmers color ad for the Model C and B was headlined "Pay Dirt Farming" and was aimed at small row crop farmers who wanted an economical cultivating tractor. One photo shows a Model C weeding a young corn crop. This ad was typical of the color ads that Allis-Chalmers placed for the next 10 years.

Look AHEAD
with Allis-Chalmers FRONT-MOUNTED Tractor Implements

Model C Self-Greasing Tractor with HYDRAULIC LIFT and DUAL DEPTH CONTROL

Sealed Reservoir bearings keep tractor automatically greased. Implements are hydraulically operated. Dual control accurately gauges depth of right and left gangs...independently. (Above—Rotary Hoe cultivator attachment.)

(Below) MODEL WC TRACTOR WITH "TOE-TIP" POWER LIFT

A touch of your toe operates "live" power lift, raising implements while in motion or standing still. To attach cultivator: Drive in. Drop two bolts and cotter key. Go!

The logical trend of cultivator and planter design has been *forward*. When implements are mounted forward on the tractor, rather than pulled behind, they need no wheels of their own. They cost less. Furthermore, they can be attached quickly and controlled automatically by tractor power.

A-C FRONT-MOUNTED planters, cultivators, fertilizer attachments and rotary hoes represent still another step forward . . . *ahead of the driver's seat.*

Front-mounted implements are brought within your natural line of vision. You watch the row ahead easily and naturally . . . never twisting to look behind.

Forward vision is an invaluable feature for precision planting, fertilizing and cultivating young plants. You'll find it not only *pays* to "look ahead"—it's a *pleasure*. Front-mounted implements are a new adventure in comfort . . . a promise of Better Living to come for the family farm.

ALLIS-CHALMERS
TRACTOR DIVISION — MILWAUKEE 1, U. S. A.

This 1947 three-color ad featured a Model C with Allis-Chalmers new front mounted implements for Model C and WC row crop tractors. "Look Ahead" was bannered on the ad showing the farmer with a clearer view of the crop with this new arrangement.

By 1960 Allis-Chalmers advertising was more graphic and detailed, as this as for the new Model D shows. The Model D was the start of a new line of tractors for the company with more convenient controls, power steering and a traction booster system. The Model D stayed in production for ten years.

BIG STICK MAKES YOU THE BOSS

Farmers who've tried other tractors can speak with authority. We asked 553 new owners, among the thousands who switched to Allis-Chalmers tractors the past year, what features they liked best.

Their answers tell you what's really new in tractors today.

No. 1 feature: Allis-Chalmers Power Director—the "Big Stick." One easy-shift lever controls 8 smoothly graduated speeds ahead in 2 ranges. Rugged oil clutch eases through tough spots with live PTO—or gives an instant surge of power when it's needed. Makes *you* the master—whatever the crop or soil.

Power Director teamed with TRACTION BOOSTER system (now with *new wider range*) matches power, speeds and automatic traction to every load.

Take a Dynamic D into the field. Test new Power Steering* . . . solid-comfort seat . . . step-on platform . . . SNAP-COUPLER hitch. Begin a new decade of productive farming—easier too—for you!

*Optional on D-14 and D-17 Models.

'60's-new dynamic D's have it!

TRACTION BOOSTER and SNAP-COUPLER are Allis-Chalmers trademarks.

ALLIS-CHALMERS, FARM EQUIPMENT DIVISION, MILWAUKEE 1, WISCONSIN

ALLIS-CHALMERS

Ask your dealer about the Allis-Chalmers plan to finance your time purchase of farm equipment.

CHAPTER THREE

J. I. CASE

Jerome Increase Case was born in Williamstown, New York, in 1819 and grew up helping his father in the operation and selling of threshers. His industrious nature served him well—by the age of 23 he had left home with six threshers in tow, determined to start his own business in Racine, Wisconsin. Quality and innovation were his hallmarks as each new thresher was introduced to the market.

His business, the J. I. Case Company, became a partnership in 1863 when Case was joined by Massena Ershine, Robert Baker, and his brother-in-law, Stephen Bull. Jerome Case died in 1891 and was succeeded by Stephen Bull as president.

With the development of the Case "Eclipse" thresher, the company became the largest manufacturer of threshers in the world—and the largest employer in Racine. Within a few years, J. I. Case was producing its own harvesters and working on improvements in mechanical power.

In 1892, the company produced its first tractor. In 1910, under the direction of Frank Bull, Stephen's son, the Case company began building luxury automobiles as the popularity of Case's steam traction engines boomed.

Efficient gas engines were the next wave. The Model 12-20 was advertised as "the best and most practical small tractor on the market." It sold for $1,095 and, like many other brands, started on gasoline and switched over to kerosene.

J. I. Case's classic 1921 ad "Keep The Boy In School" was the lead in for the Case Kerosene Tractor ad. The suggestion was that the purchase of a kerosene tractor would get the spring farm work done in half the time.

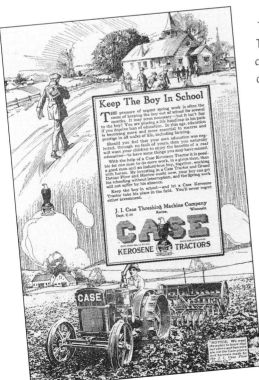

This 1921 ad from J. I. Case for their Kerosene Tractors is a classic piece of 1920s poignant ad copy. Its point was to remind the farmer that he could still achieve success on his farm and keep his children in school at planting and harvest time . . . if he owned a Case tractor.

In 1921 Case pushed their quality and durability with ads like this, informing the farmer of the Bull Gears that were in every Case tractor. The ad features a great piece of line art drawn from the activities at the Case factory.

Therefore, the help of the farmer's children was not so crucial and they could stay in school.

During this period Case also underscored its strength with an ad for Case Tractor Bull Gears which emphasized quality and endurance in every facet of production of the Case tractors claiming, "If there is going to be any breakage, it occurs in the factory—not after it is on the tractor working in the field."

Under its fifth president, Leon R. Clausen, the direction of the company took another turn. Production of steam traction engines and automobiles ceased and emphasis was placed on filling the farmers' demands for a more all-around farming tractor.

The Case 12-20 was in production from 1922 to 1928 and was a replacement for the three-wheeled model. The Model A was the next generation as row-crop specific designs were now preferred. Heavyweight models were cut out as designs became directed toward tractors that could plow, cultivate, plant, mow, and do related work with assorted attachments.

At about this time a marketing problem was solved. J. I. Case Threshing Machine Company had named another of its companies the J. I. Case Plow Works Company. Although this company had merged with the Wallis Tractor Company, the Case name was retained. Customers were often confused as to which Case company they were actually dealing with. However, in 1928 the Canadian company Massey-Harris bought out J. I. Case Plow Works

and sold the name back to J. I. Case Threshing Machine Company for $700,000. The sale ended the confusion over two companies with the same founder in the same business.

J. I. Case acquired the Rock Island Plow Company in 1937 and within two years Case's identifying color of gray enamel was converted to a livelier Flambeau Red on all of their equipment. The Model D was the first tractor to wear this new color.

The Case product line expanded, the heavy crossmount designs were discontinued in favor of standard tread tractors for row-crops, and the company became known for its versatility in machines. Models like the CC offered outstanding optional features. Its attachments, although clumsy to maneuver, could convert the tractor from a row-cropper to a cultivator; the standard axle could be set at two widths. The CC was advertised as "2 tractors in 1."

In 1942, Case produced a special Centennial plow advertised with a colorful movie called *To Freedom and Abundance with America's Finest Farm Machines*, which was shown at dealerships. The movie followed 50 centuries of man's struggle to produce sufficient food, culminating with the wonders of the modern Case tractors of the day.

Throughout World War II, Case continued to introduce new models in addition to manufacturing thousands of 155mm shells and other wartime products.

Postwar ideas included the Eagle Hitch in 1950, which made switching between

implements quick and easy. Case claimed, "Long a dream, today it comes true: Live power take-off helps save grain, makes fuel go far. Live hydraulic control is ready to act at any instant, moving or standing, when the engine is running . . . on both mounted and trailed implements."

In 1953, Case introduced the Model 500, considered to be one of the best diesel tractors on the market. This six-cylinder diesel, which came standard with dual disc brakes, electric starting, and electric lights, came with a color change to Desert Sand enamel, two-toned with Flambeau Red.

Case also released a 300 series with a new system of overlapping gear speeds. "Plow 2-3/4 extra acres a day at rock-bottom economy," Case said. The Case 300 offered a creeper gear at 1-1/2 mph, two road speeds of 12 and 20 mph, and three reverse speeds. "A gear speed to suit any crop situation."

Four years later the purchase of the American Tractor Corporation of Indiana brought Case into the production of crawler-type tractors. In the early 1960s four-wheel drive tractors were added to its line of eight different models, available with standard or sliding gear transmissions.

In 1967, Case was purchased by the Kern County Land Company, which was then taken over by Tenneco Inc., a Houston-based oil company. In 1984, Tenneco took over International Harvester's tractors and implements, making Case-IH its largest division.

A Low-Priced Tractor

IF YOU are done with experimenting—if you know now what a tractor should do for you and within what cost per year—you are ready for a Case tractor, adapted to Coast conditions.

One of the first to adopt fully enclosed construction, Case has carried this principle to the limit. Not a single working part or bearing is exposed to the action of dust and dirt. Pistons and cylinder walls are protected by an efficient air cleaner.

Only the best and longest wearing materials are used, few replacements are needed, and no expensive ones.

Case tractors hold many records for fuel economy. Operating cost is low.

No tractor is easier to handle or better adapted to work in crowded spaces.

Ample power is provided to work fast in emergencies; to work steadily day after day; to avoid overtaxing or overheating at any time.

Because of these qualities, the Case tractor lasts a long time and does a lot of work. That makes it a satisfactory tractor—one you'll always like to use. It also makes it a low priced tractor—perhaps the cheapest you can buy.

J. I. Case Threshing Machine Co.
Incorporated

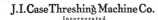

FACTORY BRANCHES: San Francisco, 235 15th St.; Los Angeles, 164-8 N. Los Angeles St.; Fresno, Calif.

Established 1842

For eighty-four years this Company has specialized on machines that made more money for farmers. The Case thresher, tractor and combine, are all examples of the high earning capacity standards of Case products.

"A Low-Priced Tractor" was the kind of line any farmer wanted to read. By 1926 the Case Company had been in the farm equipment business for 84 years. This ad focused on orchard farmers in the California Central Valley.

Headlined with "This One-Piece Main Frame," this tractor celebrated the use of a single chassis component which supported not only the lower part of the main bearings of the engine but also the front suspension, transmission, and front axle. This design was the forerunner of all modern farm tractors.

This 1938 ad for the Case Model CC featured rubber tires for the first time. This model won first place at the Wheatland Contest in the economy test, pulling a Centennial plow and turning more ground on a gallon of fuel than any other outfit. The CC could be converted from a row crop tractor to a cultivator by bolting on cultivation attachments.

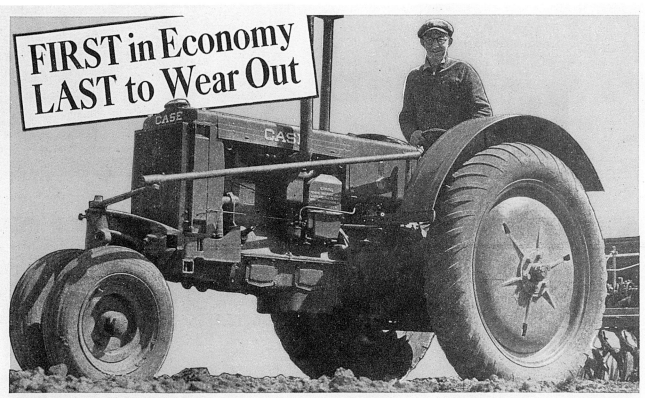

FIRST in Economy LAST to Wear Out

Today's Greatest Tractor Value at
NEW LOW PRICES!

When you buy a tractor, get the one that does MORE, for LESS money, and lasts LONGER. Get the fuel saving that won first place in the economy test at the world's foremost plowing match. At the 1938 Wheatland contest a Case tractor and Centennial plow turned more ground on a gallon of low-cost fuel than any other outfit burning any fuel at any price.

Get the tractor with the longest working life . . . that gives you more years of service for your money . . . that you can own for the lowest cost per acre or per year. Case tractors already ten years old are still going strong, still good for many more years.

Get the tractor with the lowest upkeep in tractor history. Hundreds of actual records on Case tractors five to nine years old reveal repair costs averaging only about one cent

a working hour . . . not only on gasoline, but on a great variety of low-cost fuels.

And now you can get the most economical tractor in America at reduced prices. Find out from your own Case dealer how much you save . . . how little it costs to own a Case tractor at the new *delivered* prices.

Get yourself a Case tractor now. But don't buy it on the basis of low price alone. Buy it because it brings you farm power at the lowest total cost in tractor history. Buy it because only for a Case can you get the famous "easy on—easy off" planters, cultivators, etc. Buy it to do your farming faster . . . to plow, plant, till and harvest more promptly and produce better, more profitable crops. Start right now to enjoy fast work and rock-bottom economy with a Case tractor. Mail the coupon today.

Save Power, Time and Money With Today's Lightest-Draft Plow

Plow faster with less fuel. Get the plow that helped win the economy record at Wheatland. The Case Centennial Plow saves power that ordinary plows waste in landside friction; easy adjustment to carry landside pressure on the wheels. A wonder in weeds and trash because it has clearance where clearance counts; high-speed bottoms cover clean at all depths. Higher, quicker level lift, with *power clutch* in *oil bath*. Sizes to suit all farms and all tractors.

SEE What You Get in a CASE

Power-Saving Transmission
Synchronized Steering
Sleeve-Type Cylinders

Oil-Bath Air Cleaner
Complete Dust Protection
Full-Pressure Lubrication

IT COSTS LESS TO FARM WITH
CASE

In 1948, this two-color Case ad offered 20 great models, also noting the company's presence in power machinery for the last 106 years.

The Case DC was introduced in 1939 and remained one of the most popular models until 1955. This 1946 two-color ad asks, "Have you been wanting more Power?"

The new Case Diesel 500 was powered by a six-cylinder "Powercel" diesel engine. Case claimed this was a fast-stepping five-plow tractor. With the "500" came a color change to Desert Sand and Flambeau Red.

STAR OF ALL DIESELS

MODEL "500"

NEW CASE DIESEL

FIRST WITH POWER STEERING

★ "Powrcel" Controlled Combustion ★ Direct Electric Starting with Diesel Fuel ★ Six-Point Fuel Filtering System ★ Single-Plunger Injection Pump ★ 5-Plow Power ★ Dual-Valve Constant Hydraulic Control ★ Constant Power Take-Off ★ 7-Bearing Crankshaft ★ 6-Cylinder Engine

Save precious hours . . . keep labor and production costs down, yields up . . . with this big, fast-stepping 5-plow tractor—the new Case Diesel. Enjoy the ease of Power Steering for short turns and long days, especially in soft or rough ground. Get smooth, quiet power and clean burning at all loads with "Powrcel" controlled combustion. Thrill to instant starting directly on Diesel fuel by merely touching a button. See how the 6-point filter system protects against dirt or water in the fuel . . . how the single-plunger injection pump with distributor feeds fuel evenly to all six cylinders. Best of all, you'll have economies in maintenance never before achieved in a Diesel tractor . . . plus the benefit of Diesel fuel economy . . . and many other features of advanced design.

See your Case dealer now for a personal demonstration in your own fields. Ask him about big, new Case plows, disks and drills that match the power of this star of all Diesels.

SEND FOR THE FULL STORY

Get all the facts and pictures. Mark here or write in margin any size tractor, any kind of implement or farm machine that interests you. Send to J. I. Case Co., Dept. C-774, Racine, Wis.

☐ Case Diesel Tractor ☐ 5-bottom Case
☐ 18-ft. One-Way Plow Centennial Plow
☐ New Wheel-Type ☐ New Lister-Type
Tandem Disk Harrow Press Drill

Name_____

Post Office_____

RFD_____ State_____

CASE

"THE BEST CROP OF OATS I EVER RAISED"

"The Case grain drill I purchased is almost human. I set it for 10 pounds of alfalfa seed and it sowed exactly 10 pounds per acre. The best crop of oats I ever raised was this year under unfavorable weather conditions—67½ bushels per acre, much better than any of my neighbors, and which I attribute to the Case drill.'—*Harvey Laqua.*

THANKS TO THE ACCURACY OF THE
SEEDMETER
IN CASE DRILLS

PADDLE-WHEEL AGITATOR · LOW HOPPER · NEW ROLLER BEARINGS

Low and large, hoppers are easy to fill, make long rounds between fills. Strong steel frame resists sagging, keeps working parts in line, prolongs the accurate life of drill.

Revolving paddles keep bearded oats and other bulky, rough seeds from bridging, assure steady supply of seed at Seedmeter. Available as extra equipment, not needed for most standard crops.

Roller bearings are available at extra cost for single-disk furrow openers. They reduce friction, help disks turn freely, work better. Need lubrication only occasionally —save much time.

● **There's more difference in grain drills** than you might think. To get full yields of grain, full stands of grass, you need the right amount of seed in every foot of every furrow. The exact shape of the shallow-fluted rolls, the deep seed cup with adjustable gate, the wide revolving ring, the two-speed drive—all play their part in the amazing accuracy for which Seedmeter is famous. Grass-seed attachment is built with Seedmeter in smaller size. See your Case dealer now about type, size, spacing and furrow openers to fit your "farming. Plan now for better stands, cleaner fields and higher yields through the years ahead.

SUCCESSFUL FARMING, FEBRUARY, 1951

CASE

PASTE ON PENNY POSTCARD AND MAIL

Get latest folders. Mark machines that interest you; write in margin any others you may need. J. I. Case Co., Dept. B-77, Racine, Wis.

☐ Fertilizer drills   ☐ Low-cost "VAC" tractor
☐ Plain drills   ☐ Larger 2-plow "SC"
☐ Disk harrows   ☐ 3-plow "DC" tractor
☐ Tractor plows   ☐ 4-5 plow "LA" tractor

NAME _____

POSTOFFICE _____

RFD _____ STATE _____

11

Case offered not only tractors but a wide variety of farm implements including this grain drill which incorporated a seed meter. It was endorsed by Harvey Laqua, a farmer who had gained 67 1/2 bushels per acre by using the machine.

In this 1957 ad for the Case 350 the color change to Desert Sand and Flambeau Red can be seen. The ad features a beautiful colored illustration; this ad was one of a series of similar ads from Case.

Announcing...
Great New 3-4 Plow
More Power...
More Weight...
Latest
More Traction...

NOW you can push your work faster . . . and better . . . than ever before was possible with implements of 3-4 plow rating. The new Case 350 gives you a torrent of 41 horsepower from its big 164-inch, high-compression engine, perfectly geared to every task with its 12-speed Tripl-Range transmission. The row-crop models have 36" wheels and 25" clearance to cultivate tall row crops. The standard 4-wheel model is lower for close work in orchards or groves. Both models combine length and weight for excellent stability and traction. With constant-running PTO and plenty of reserve engine power, the 350 works wonders on harvest jobs. Break through the 3-plow barrier . . . crowd toward 4-plow capacity . . . with this latest, greatest Case tractor. See it, first chance, at your Case dealer's.

Watch for your Dealer's Caseorama Starlight Review featuring the new, broader than ever Case 12-month line starring new wheel and crawler tractors for every crop and utility requirement on and off the farm. Watch for the date, mark your calendar, be sure to be there.

By the mid-1950s, Case's ads showed the perfect farm with the perfect farmer driving a spotless Case 300. Case claimed that this tractor, with its faster fourth gear and a 4-14 plow, could cover four and a half more acres per day.

CHAPTER FOUR
JOHN DEERE

One of the most enduring companies in the farm tractor business was founded by John Deere. He was born in Rutland, Vermont, in 1804 and became a blacksmith's apprentice, opening his own blacksmithing business in Grand Detour, Illinois, in 1837. His shop flourished as he moved from general repair to fabrication of farm implements. He recognized a need for a better plow and with a keen eye and creative hands, built it, and within a few years John Deere became a major manufacturer of horse-drawn plows and seed drills.

With the help of a new partner, Leonard Andrus, Deere moved ahead with his plans to expand his plow-making facility. He discovered that the center for this type of manufacturing lay in Moline, Illinois, and around 1850 moved his business to Moline, where he split from Andrus. Over the next ten years, he expanded his business, buying up several smaller companies. Following John's death in 1886, his son Charles took over and expanded the company even further. One popular product was the Gilpin Sulky plow, invented by Gilpin Moore, Deere's shop manager.

It wasn't until 1912 that Deere & Company began experimenting with powered tractors. After six years of testing, however, they still did not have a production model. In March 1918, they purchased the Waterloo Gasoline Engine Company, makers of the Waterloo Boy tractor, and this purchase instantly gave them a line of highly successful tractors from which they could develop their own tractors.

It is obvious from Deere's print advertising that they were proud of their new line of Waterloo Boy tractors. Deere's advertising in 1920 trumpeted the 25 horsepower Waterloo

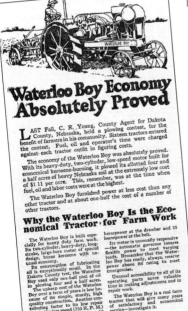

In August 1921, county agent C. R. Young (Dakota County, Nebraska) held a plowing contest for the benefit of farmers in his community. This ad touted the Waterloo Boy's winning low cost of $1.11 per acre to operate.

This 1928 ad for the first John Deere tractors, Model D, had crisp display line art which made it stand out on a newspaper page. It was hoped that the banner headline of "Its Best Salesmen are the Men who use It" drew farmers' attention to its benefits.

Boy for its economy, powered by a two-cylinder engine that "burns kerosene completely." This was part of Deere's extensive ad campaign for these tractors, denoting that they were the "heavy-duty type" and that they used only "straight cut gears because beveled gears cause friction." The ad was laid out in line art like a dream farm scene with a meticulous farmer disking perfectly flat land under a sky touched with fluffy white clouds.

By 1920, the company was offering an extensive line of farm machinery products, including binders, cultivators, harrows, hay loaders, mowers, wagons, and tractors. In 1923, the first John Deere tractor, the Model D, was introduced followed by the smaller Model C in 1927. In its 1928 advertising Deere maintained a low key, noting that, "Its Best Salesmen Are the Men Who Use It" and that you could write for a free booklet titled *What the Neighbors Say,* which contained recommendations and endorsements from current Deere tractor users.

It was with the Model D that Deere made its name in farm tractors, keeping mutations of the Model D in production for almost 30 years. One of Deere's first tricycle-wheeled tractors was the GPWT (Wide-Tread) offered in 1929.

John Deere jumped into the tractor business overnight when the Waterloo Gasoline Engine Company was purchased in 1918. This January 1920 ad for the two-cylinder kerosene-powered 25 horsepower Waterloo Boy notes that "Waterloo Boy gives you economical power."

The rugged power plant of the Waterloo Boy Tractor. Note the compact, clean-cut design. The reliability and economy of this engine has made the Waterloo Boy popular with farmers all over the country.

Why a Waterloo Boy Gives You Economical Power

THE Waterloo Boy Tractor has built among its many owners a reputation for steady, dependable power, and economical, money-making service. Back of this power and service stands its sturdy two-cylinder engine.

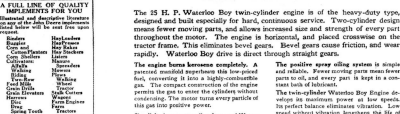

WATERLOO BOY
BURNS KEROSENE COMPLETELY

The 25 H. P. Waterloo Boy twin-cylinder engine is of the heavy-duty type, designed and built especially for hard, continuous service. Two-cylinder design means fewer moving parts, and allows increased size and strength of every part throughout the motor. The engine is horizontal, and placed crosswise on the tractor frame. This eliminates bevel gears. Bevel gears cause friction, and wear rapidly. Waterloo Boy drive is direct through straight gears.

The engine burns kerosene completely. A patented manifold superheats this low-priced fuel, converting it into a highly-combustible gas. The compact construction of the engine permits the gas to enter the cylinders without condensing. The motor turns every particle of this gas into positive power.

Simplicity is an outstanding feature of Waterloo Boy design. You don't have to be an expert to care for it. There are a number of inspection plates. Each one is conveniently placed. Every part is easy to get at from a standing position. Any adjustment easily made.

The positive spray oiling system is simple and reliable. Fewer moving parts mean fewer parts to oil, and every part is kept in a constant bath of lubricant.

The twin-cylinder Waterloo Boy Engine develops its maximum power at low speeds. Its perfect balance eliminates vibration. Low speed without vibration lengthens the life of the motor and reduces the upkeep.

The real test of a tractor is field performance. The Waterloo Boy engine has high field efficiency. It is giving thousands of satisfied owners dependable, economical power.

The purchase of a tractor is an important investment. Investigate thoroughly before you buy. We have a booklet fully describing the Waterloo Boy and its powerful engine. Don't fail to get it. Drop us a postal today. Address JOHN DEERE, Moline, Illinois, and ask for Booklet W B 29.

A FULL LINE OF QUALITY IMPLEMENTS FOR YOU

Illustrated and descriptive literature on any of the John Deere implements listed below will be sent free upon request.

Binders	HayLoaders
Buggies	HayPresses
Corn and	Hay Rakes
Cotton Planters	Hay Stackers
Corn Shellers	Listers
Cultivators:	Manure
Alfalfa	Spreaders
Walking	Mowers
Riding	Plows
Two-Row	Walking
Feed Mills	Wheel
Grain Drills	Tractor
Grain Elevators	Stalk Cutters
Harrows	Farm Engines
Disc	Farm
Drag	Tractors
Spring Tooth	

John Deere Implements are distributed from all important trading centers. Sold by John Deere dealers everywhere.

THE TRADE MARK OF QUALITY MADE FAMOUS BY GOOD IMPLEMENTS

This machine helped the introduction of two- and four-rowed implements.

The Model A, first built in 1934, turned out to be one of the most popular of all Deere's tractors and was built in various forms until 1954. The Model B, a smaller version of the A, was also offered. The Model B was heralded as "the small tractor that farmers demanded" and was simply a two-thirds scale version of the Model A. Both these tractors offered adjustable tracks, four-speed transmissions, and simplicity of operation.

During the 1920s and early 1930s, many two-color ads were displayed by tractor manufacturers, but Deere was one of the first to advertise in full color. This was very limited initially, but by the beginning of the 1940s, when other tractor manufacturers were first beginning to use color, it was easy to tell who had the budget to make the big splash with the new four-color printing.

This move to color ads was a distinct change in direction for Deere, which had previously, like all the others, only advertised in black and white or two-color using line art and illustrations. One ad noted that Deere had 14 distinct models on the market and its 1939 ad for the Model H proclaimed, "You'll be Mighty Glad You Bought a John Deere."

On through the 1940s and into the 1950s, Deere continued as king of the two-cylinder tractor, advertising not so much specific models but the John Deere brand in general.

The AO, the orchard version of the Model A, was a highly popular product from Deere and had a

typically long production run. Deere's first diesel arrived on the market in 1949 as the R model, but it only survived four years.

The John Deere Model D tractor weighed 4,000 pounds but offered 15 horsepower at the drawbar and 27 at the belt pulley. The ad said, "Performance is What Counts."

A crawler-style tractor was also offered. The conversions were done by the Lindemann Company in Yakima, Washington. This unit used a Deere BO chassis mounted to one of Lindemann's crawler units. Deere then built its own versions including the 40C, which stayed in production from 1953 to 1955.

Around this time the Deere Company changed its model designation from alphabetic to numeric, and all future models were titled with model numbers like the 70 series or the 435 series. During the latter part of the 1950s numerous new models were introduced: the 20, 30, 320, 420, 520, and 820. Some models were truly new, while others were different only in sheet metal.

John Deere's advertising expressed "Quality Farm Equipment" with products that were economic to use and low priced. They noted that "John Deere was the Choice of the Tractor-Wise."

It is noticeable that in the 1950s, John Deere liked having independent recommendations for their products as it regularly featured farmers who "loved" their Deere tractors. These ads noted, "Planted at high speed with less work," "A more accurate planter I've never seen," and "The one-man haying system saves me $20 a day." All of these were great endorsements and no doubt spoke right to the heart of farmers looking to purchase new machinery.

An ongoing line of John Deere tractors in both gasoline and diesel configurations has been offered since the early 1950s, plus four-wheel drives, automatic transmissions, articulated chassis, and special-purpose tractors. Today, John Deere is still considered to be the builder of some of the best farm tractors in the world.

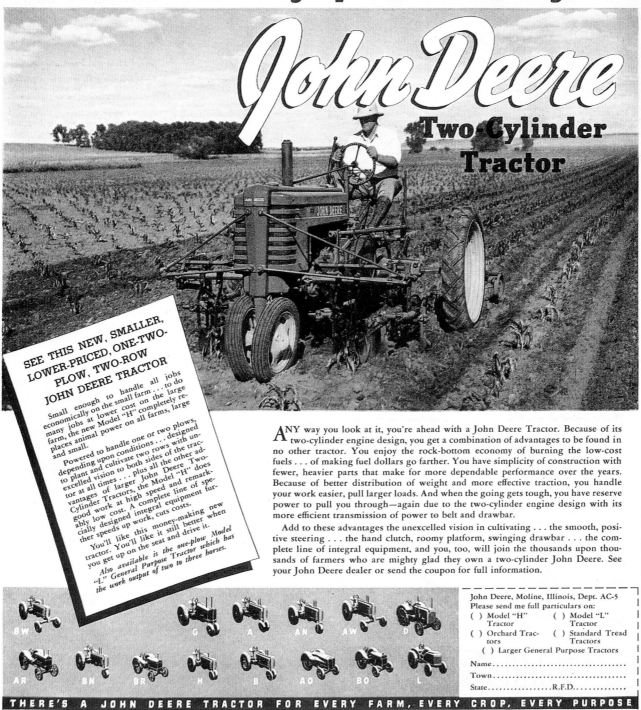

This December 1939 John Deere ad was among the first four-color full-page tractor ads run by the company. It featured 14 different Deere models with the Model H weeding a cornfield. It was bannered with, "You'll be Mighty Glad you Bought a John Deere."

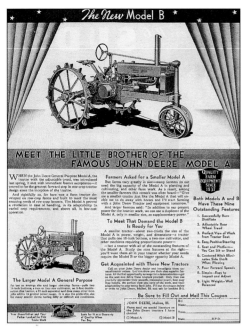

Deere's first tricycle tractor arrived in 1929 with the GP Wide-Tread model. By 1935, the A and B models had replaced it and offered even more outstanding features. The new model B was presented in the ad like a stage celebrity.

In October of 1949 "again John Deere steps out ahead" with five new models including the OA Orchard, MC Track, and R Model diesel. The ad was printed in a strange four-color process which mixed illustration with photography.

This mid-1950s ad for John Deere's new 420 fleet was for a range of 10 highly specialized tractors including crawlers, tricycle chassis versions, and high-wheel models. This full-color ad used a photo illustration and was bannered "Powerfully Good News."

Choice of the Tractor-Wise

WHEN it comes to determining tractor value, "experience is the best teacher." Thousands of today's John Deere owners know from personal experience with other tractors that you just can't equal a John Deere. These tractor-wise farmers, and thousands more who "value-shopped" before they bought, have found that the exclusive John Deere "two-cylinder idea" really pays off in more dependable performance season after season . . . in fewer and far lower repair bills down through the years . . . in longer tractor life.

Equally important, these owners have found that John Deere's advanced engineering pro-

vides a greater combination of modern operating features to speed up every power job, do it better, make it easier.

The more you know about John Deere Two-Cylinder Tractors, the more convinced you'll be that a John Deere is the tractor for you. See your John Deere dealer for the complete facts and a demonstration of the size and type that fits your needs. Compare it on every count with any other tractor you could own. We feel certain you'll be on your way to more profitable, more enjoyable farming—with a John Deere. For free literature, fill out and mail the coupon below.

JOHN DEERE

Moline, Illinois

John Deere, Moline, Illinois   Dept. D8

Gentlemen:
Please send me free literature on John Deere General-Purpose Tractors.

Name...
R.R......................................Box No..........................
Town...............................State..........................

John Deere also ran a lot of generic tractor advertising promoting the company's image as the confident builder of fine machinery. This full color ad again shows the perfect tractor plowing the perfect corn crop.

...40 50 60 and NOW the 70

POWERED for the large row-crop farm...
DESIGNED with every modern tractor feature

MEET the Model "70"—a brand-new John Deere general-purpose tractor *with more than fifty* horsepower!*

Here is *pulling power* to handle four 14-inch plow bottoms and five 14's in many soils—power to work 4-row bedders . . . 12- and 14-foot double-action disk harrows . . . 20-foot disk tillers and other large tillage tools at maximum capacity and big savings in time and labor.

Here is *"live" hydraulic power*—through John Deere's famous Powr-Trol—to raise, lower, and adjust all your integral and drawn machines and make control of your equipment easier and faster than ever before.

Here is *"take-off"* power to save you the cost of purchasing and maintaining auxiliary engines for your driven machines. You can choose either the transmission-driven power shaft or the "live" power shaft that will still further speed up your work and save you extra clutching and shifting.

Here is *heavy-duty row-crop power*—in your choice of interchangeable front-wheel equipment—available with a complete line of high-quality, big-capacity integral tools to

do your tilling, planting, cultivating, loading, corn picking and other jobs just the way you've always wanted to do them.

Here is *truly modern power*—with six forward speeds and all the advanced engineering features introduced and field-proved on the Models "50" and "60" Tractors, which have been enthusiastically acclaimed by farmers everywhere.

Here—in the new Model "70"—is *More Power* to do *More Work* on *More Jobs* than in any other tractor of its size and type. Your John Deere dealer is ready to prove it to you in the field. Ask him for a demonstration.

* Belt: 50.35; drawbar: 44.21. Sea level (calculated); maximum h.p. based on 60° F. and 29.92 in. Hg. Figures given are for gasoline engine; all-fuel engine optional at no extra cost.

JOHN DEERE, MOLINE, ILL.   Dept. C25

Please send me free folder on the new John Deere Model "70" Tractor.

Name..

R. R................................Box........................

Town.........................State...................

JOHN DEERE TRACTORS

In 1953 John Deere introduced the Model 70, a larger row crop tractor with more than 50 horsepower. The full color ad centered around a traditional full-color illustration.

Deere's 1960 ad featuring the Model 730 Diesel Tractor pulling two four-row planters and a two-planter hitch in a cornfield in the Midwest. The planter was endorsed by four farmers who all found it to their benefit, including the Jackie Gleason look-alike, Virgil Zimmerman from Scranton, Iowa, who claimed the 494 Planter offered "accurate trouble-free planting."

Here's the promise...

There's no better way under the sun to make QUALITY HAY at low cost...JOHN DEERE Hay Tools are your best buy

Hay a cash crop? You bet it is! It just takes a little processing to become meat or milk. The better the hay, the more meat or milk you produce. Give your hay crops preferred treatment with John Deere equipment and you'll get cash results in your dairy or livestock checks.

Everybody Knows . . .

It's no secret that the leaves are the "meat" in your hay so why settle for stems or "bones" of your crops? Mow 35 or more acres per day with a John Deere Mower to cut your crops at their peak in feed value. Condition your crops with a John Deere Hay Conditioner—its crimping action cuts curing time in half, reducing weather risks. There's two-way profit in a mower-hay conditioner combination—you'll save time and money in a once-over field operation.

Aerating your hay with John Deere's new, low-cost Swath Fluffer speeds curing—saves rained-on crops, too. Handle your hay with "kid gloves" with the gentle action of a John Deere Rake. You'll make loose, leafy, faster-curing windrows.

Last but Not Least, . . .

Add the final touch of a dependable, big-capacity John Deere Baler. You'll put up your hay faster to pack the maximum feed value into every bale. You can even turn baling . . . loading . . . and storing into a one-man show with a Bale Ejector for all John Deere Twine-Tie Balers and the back-saving Elevator-Barn Conveyor Combination.

Yes, sir, it's true—you do make better hay the John Deere Way. See your dealer soon.

John Deere Balers are built simply to make your baling operations faster . . . easier . . . lower cost.

Here's the proof...

My John Deere 14-T Baler Keeps Six Men Busy—Has Held Upkeep Costs to $15 in Four Years' Operation

"I have been very well pleased with my John Deere 14-T Twine-Tie Baler for the last four years. The 14-T has the capacity to keep six men busy—one with the tractor and baler, two on the wagon, and three in the barn. One day it turned out 1,200 bales in less than 5 hours. We have baled 45,000 bales and the upkeep costs on the 14-T were $15.00. I highly recommend John Deere Hay Equipment to anyone who wants to do the job at the lowest possible cost."

Richard H. Eltzroth, Wabash, Indiana

"Our No. 8 Mower-Hay Conditioner combination enables us to handle two jobs with one man. The mower is two years old and the repair costs have been about $5. The conditioner cuts our curing time by a day . . . saves leaves . . . and makes greener colored hay that isn't wasted by the cattle."

Richard Uhlenhake, Burlington, Wis.

"The John Deere One-Man Haying System saves me $20 a day every day I'm baling hay. The 14-T Baler with Bale Ejector and the Barn Conveyor save the work of two men. We're also making better hay because we put it up faster. I figure it makes haying so easy that I can now farm for another 10 years."

M. Ulferts, Hammond, Wis.

"My 214-W Wire-Tie Baler is a tremendous improvement over any other baler because of its big capacity and low maintenance. I've put up 100,000 bales and had no work done on it. In custom work, its big capacity and dependability make the difference in the profit."

Ralph Bodine, Stilwell, Kans.

SEND FOR FREE LITERATURE

JOHN DEERE • MOLINE, ILL. • DEPT. N38

Please send me free folders on the ☐ No. 8 Caster-Wheel Mower ☐ 3-Point-Hitch No. 9 Mower ☐ New 10 Side-Mounted Mower ☐ Hay Conditioner ☐ Swath Fluffer ☐ Rakes ☐ Twine-Tie Balers ☐ Bale Ejectors ☐ Wire-Tie Balers ☐ Bale Elevators ☐ Credit Plan.

Name_____ ☐ Student

Rural Route_____ Box_____

Town_____ State_____

JOHN DEERE
MOLINE, ILLINOIS

"WHEREVER CROPS GROW, THERE'S A GROWING DEMAND FOR JOHN DEERE FARM EQUIPMENT"

40

41

By 1960, endorsement advertising had become a great way to immediately prove to buyers that your product was superior. Once again, John Deere had four endorsements for their 14-T Twine-Tie Bailer. According to the good looking Richard Eltzroth from Walbash, Indiana, "I've been well pleased with my 14-T Twine-Tie Bailer."

At the end of the 1950s, John Deere was one of the leading manufacturers of farm equipment and tractors. This neat four-color ad asked, "What makes two work cheaper as one?" A John Deere of course, with production costs as low as $5.00 an acre.

What makes two work cheaper as one?

Put John Deere power in front of a minimum-tillage hookup and you can cut production costs as much as $5 an acre. Makes good farming sense—put more of a load on tractors . . . double-up on tools . . . get several operations done once-over.

John Deere saw the handwriting on the wall in the need for bigger tractors and put out models ideally suited to double-up operations. The 65 h.p. "3020" * and 91 h.p. "4020" * deliver power for multiple-unit hookups not only on the drawbar but through the PTO and hydraulic systems as well.

Hand-in-glove with the development of these tractors, John Deere engineered equipment-ganging devices to make big tractor power pay off. Minimum tillage hitches. Squadron hitches that link together up to 40 feet of drills. Tool carriers brawny enough to handle several types of implement attachments simultaneously.

Your John Deere dealer wants you to take a crack at farming with double-up power. He'll supply the tractor and implements if you'll supply a few hours of your time. When the trial is over, he'll make it easy to shake hands on a tractor and implement deal with good John Deere Credit Plan terms.
*Diesel model

JOHN DEERE
Moline, Illinois

John Deere double-up power!

CHAPTER FIVE

FORD

Henry Ford had his finger in every mechanical pie. The overwhelming success of his Model T allowed him to indulge in many other interests, including farming. As early as 1906 Ford was experimenting with a lightweight, low-cost farm tractor, which he called an Automobile Plow. Henry's dream to build a viable farm tractor came to fruition in 1916 when he introduced the Fordson, built by a separate company, Henry Ford & Son.

Apparently the tractor wasn't the great engineering feat that the Model T had been, but despite its failing it still sold over 34,000 units in its first year.

World War I had provided Ford with valuable contracts to supply large numbers of tractors to the British government so that they could improve British farming methods. At the war's end, Ford foresaw significant sales to industries that also needed tractor-type hauling vehicles. Industries as diverse as mining and sawmills needed tractors, and the economic benefits of owning one were clearly spelled out in Ford's ad around 1920 which featured a Fordson owned by the Julius Seidel Lumber Company in St. Louis.

Over the next eight years a price war ensued as a result of a general recession. Prices of tractors that were once thousands of dollars were lowered into the hundreds. Henry Ford, in an attempt to sell off existing inventory, kept lowering his prices, and many tractor manufacturers could not compete—they were forced to close their doors forever. The only company that had the power and resources to counter Ford was the giant International Harvester.

It was at this time that International Harvester came out with the immensely popular

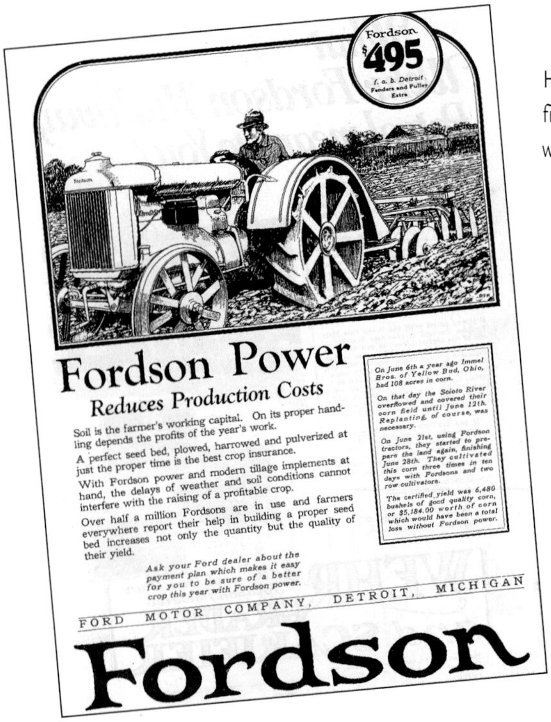

Ford's price war raged with International Harvester Company through the 1920s. In this fine line art ad from 1926, the Fordson tractor was offered at $495. Before Ford quit the market he had dropped the price another $100. However, Ford still could not compete.

Henry Ford saw alternative uses for his tractors as this 1923 ad for Fordsons, owned by the Julius Seidel Lumber Company in St. Louis, displays.

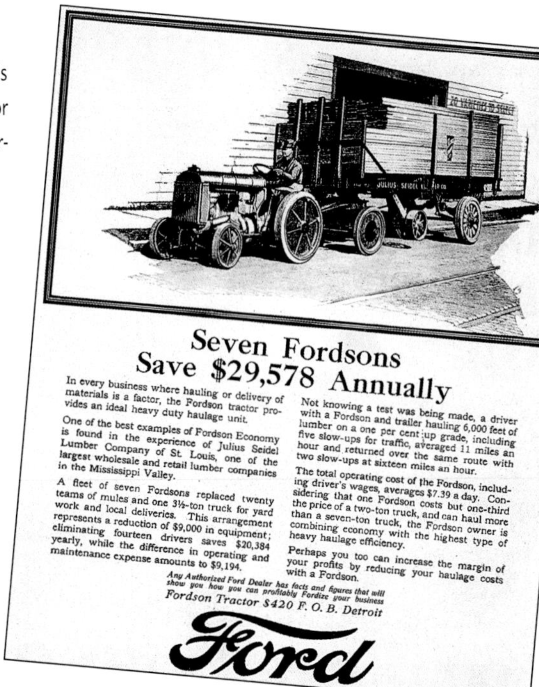

and much-loved Farmall. The Fordson was a great little tractor, but it couldn't do everything, and farmers were crying out for a more versatile tractor. The Farmall was a hit. It was great for cultivating crops, it had a power take-off, and it could replace all of the horses on the farm; whereas, the Fordson could not. Henry Ford found that even his newest Fordsons could not compete with the Farmall, and he transferred production first to Ireland, and then to England.

In the early 1930s, some Irish Fords were imported to the United States along with a few tricycle, rubber-tired English Fordsons built at Ford's Dagenham automobile factory on the outskirts of London, England. In 1938 the Ford tractor scene changed for the better when Harry Ferguson agreed to let Ford use his system of a hydraulically powered, three-point linkage in the new Ford 9N tractor. The 9N was succeeded in 1942 by the 2N, which was produced until 1947.

The improved 8N model followed, as did the purchase of Wood Brothers Inc., a farm implements manufacturer. Ford and Ferguson eventually had a falling out and their verbal agreement, based on a handshake, ended unhappily in 1946.

Ferguson went on to sell his own line of tractors and Ford regrouped with the Dearborn Motors Company, which acted as the manufacturer and distributor of Ford tractors and implements.

In 1952 the English-built Fordson Major rolled onto the market. This well-built Ford was manufactured at Ford's Dagenham plant and was quite a hit in England, the United States, and the rest of the world, with its bannered advertising pointing out its great money-saving and money-making abilities.

In recognition of its fiftieth anniversary, Ford introduced the "Golden Jubilee" NAA in 1953 and resumed distribution of Ford farm tractors and equipment. The new Ford Workmaster and Powermaster models were part of the 500, 600, 800, and 6000 diesel series that followed over the next few years.

By the end of the 1950s Ford had swung heavily back into the tractor business, creating a diverse offering of tractors for crop farming. In 1958, the 861 Diesel was touted as the best three-plow tractor available offering 50 percent better fuel efficiency than comparable gasoline tractors, and the 961 row crop model offered a full line-up of well-designed Ford row crop implements, including a front-mounted cultivator.

In 1962, the traditional gray and red paint scheme on Ford's U.S. tractors was replaced with the blue and white long familiar on the English models.

Many of the tractors that plowed the American farms in the 1950s were Fords, and the company went on to far greater heights with the purchase of New Holland and the sale of super-sized tractors such as the Steiger four-wheel drive.

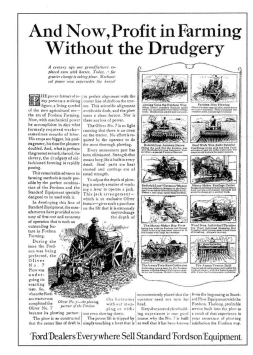

By 1922, when this ad appeared, the Fordson Tractor had been on the market for four years. Their sales program was successful even though they had encountered mechanical problems. Henry Ford's hope was to offer farmers an inexpensive tractor for their general farming operations. This ad illustrates the tractor using implements from several different implement makers including Amsco and Roderick Lean.

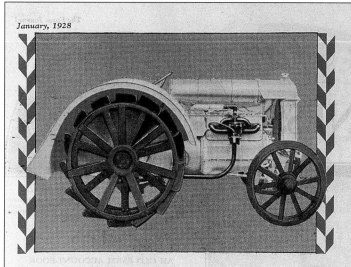

This two-color 1928 ad for the Fordson still had the price pegged at $495 but the end was near for this popular tractor that could be used for plowing, belt work, snow removal, and sawmilling.

In this 1953 ad for "The Golden Jubilee," the NAA model tractor tows a Dearborn Farm Equipment Corn Picker. This combination highlighted Elmer and Verna Bollinger's win in the National Mechanical Corn Picking Contest.

In 1936 Harry Ferguson and Henry Ford came to an agreement to create a Ford tractor using the Ferguson hydraulic linkage system. In this 1941 ad, Ford promoted 4H Clubs, Future Farmers of America, and the National Farm Youth Foundation as the way to the future, especially if Ford tractors were used.

"The best job of Machine Picking we've ever seen!"

That was the opinion of judges and spectators as 65,000 people watched Elmer Bollinger, his Ford Tractor and Dearborn-Wood Bros. Corn Picker win the National Mechanical Corn Picking Contest near Rushville, Indiana, last fall

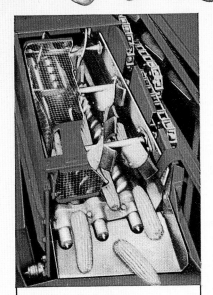

This "bed" can help you sleep easier!

This extra large husking bed has 6 instead of the usual 4 rolls—and it's the size of the husking bed that largely determines how much *clean husked* corn you can put in the wagon. You can sleep easier, knowing that your corn is free of debris, husks and silk (common causes of spoilage in the crib).

There are two important reasons why the Dearborn-Wood Bros. Corn Picker is the first choice of so many farmers. They find it *picks cleaner corn—and more of it!*

Naturally, this picker's ability to get more corn—and husk it cleaner—can help fill the cribs on your farm with extra bushels of profit, too.

So the first chance you have, watch a Dearborn-Wood Bros. Corn Picker in action. See how the picker gently noses under down stalks and gets more ears into the machine. Notice how *three* sets

National Champion Elmer Bollinger and his wife Verna, of Fairbury, Illinois beamed all over when he was awarded these two gold trophies.

of gathering chains keep the crop moving, even in down and tangled corn. Watch its exclusive rotary snapping bar team with snapping rolls to get more ears, help clean trash, reduce shelling. Then *six* long husking rolls go to work fast, stripping off husks and silks gently, yet cleanly.

So if you want your corn crop to get the best picking — profit-making, championship picking at low cost — stop in and see your nearby Ford Tractor dealer about the Dearborn-Wood Bros. Corn Picker.

Ford Farming

MEANS LESS WORK . . . MORE INCOME PER ACRE

Dearborn
FARM EQUIPMENT

DEARBORN-WOOD BROS.
CORN PICKER

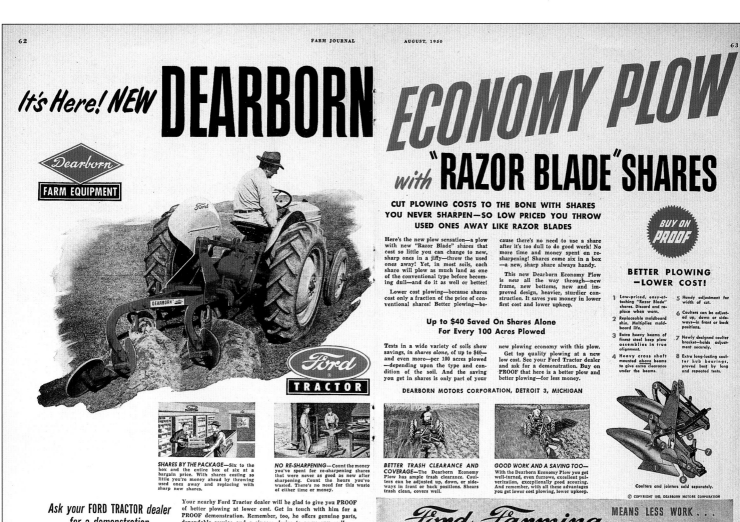

This ad for Ford's new economy plow with "Razor Blade" shares mixed illustration and photography and told the buyer to "Ask your Ford tractor dealer for a demonstration."

In 1954, Ford had 31 different ways of proving that its new 600 tractor was better than ever.

"I picked 176 acres of corn for only $14⁷⁰ !"

— says Cecil Baggs, Tabor, Iowa

"In harvesting 176 acres of corn," Mr. Baggs reports, "I picked over 6,000 bushels with my Fordson Major Diesel on 98 gallons of diesel fuel. The fuel cost was $14.70 — about 8 cents per acre."

Low cost FORDSON MAJOR DIESEL

"Handles like a small tractor but has the power of a big one. I pull four plows in third gear." B. Stalboerger, Waubun, Minn.

"In pulling a field chopper, my previous tractor used 25 gallons of leaded fuel a day. My Fordson Major uses only 10 gallons a day — fuel costs are only one-third as much." Lester Hoffbauer, Westport, South Dakota.

"Drives so easy, my neighbors thought it had power steering." Joe McTighe, Maurine, South Dakota.

"We have 3 other diesel tractors, but the Fordson Major pulls more, is easier to handle and uses less fuel." Otto Bertsch, Halstad, Minnesota.

"I have two other tractors, but don't use them much because my Fordson Major uses half the fuel. It rides and handles much easier, too." Lloyd Clow, Oreans, Minn.

TRACTOR AND IMPLEMENT DIVISION • FORD MOTOR COMPANY • BIRMINGHAM, MICHIGAN

Ford Farming

GETS MORE DONE... AT LOWER COST

Cecil Baggs from Tabor, Iowa, claimed, "I picked 176 acres of corn for only $14.70!" with his new Fordson Major diesel tractor. Lester Hoffbauer from Westport, South Dakota, said, "Fuel costs are only one-third as much." These diesel tractors were built at Ford's Dagenham plant in England and exported worldwide.

By the mid-1950s, Ford was offering over 70 implement options for the 800 series. This full color ad shows a Model 860 fitted with an 8 1/2-foot Flexo-Hitch disc harrow. The ad also noted that Ford had already sold over two and a quarter million tractors.

America's Largest

Shown here is Model 860 Ford Tractor with 8½-foot lift-type Ford Flexo-Hitch Tandem Disc Harrow

Your choice of more than 70 quick-attached mounted implements ...and 11 Ford Tractor models both Tricycle and 4-wheel

Shown here is just a handful of the *quick-attached* implements in the Ford line. Ford also offers a large number of pull-type implements including major harvesting machines.

Ford implements are made in many sizes and types to fit the 11 models of 2-plow and 3-plow Ford Tractors and also the 3-4 plow Fordson Major Diesel.

The next time you're in town see your Ford Tractor and Implement Dealer—he's a good man to know better. Tractor and Implement Division, Ford Motor Company, Birmingham, Michigan.

Ford Farming
GETS MORE DONE...AT LOWER COST

Rear Mounted Mowers

Economy Moldboard Plows

Rear Mounted Loader

Rigid Type Disc Harrow

Field Cultivator

Subsoiler

Spring Tooth Harrow

Rotary Cutter

Two-Way Plow

Over 2¼ Million Ford-Built Tractors...Over 1

Pick up and Go" Family

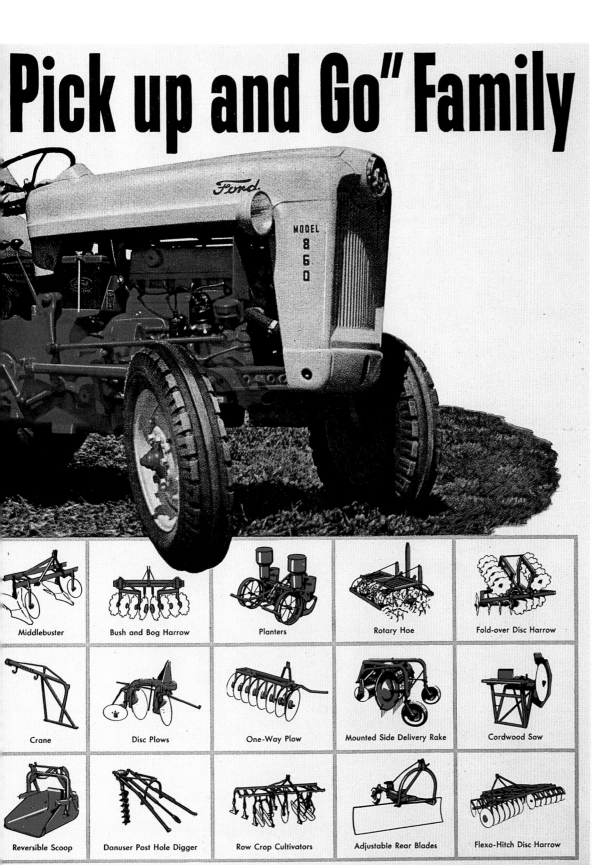

Middlebuster

Bush and Bog Harrow

Planters

Rotary Hoe

Fold-over Disc Harrow

Crane

Disc Plows

One-Way Plow

Mounted Side Delivery Rake

Cordwood Saw

Reversible Scoop

Danuser Post Hole Digger

Row Crop Cultivators

Adjustable Rear Blades

Flexo-Hitch Disc Harrow

Million of them with 3-Point Hydraulic Hitch

New model 961 row crop Ford tractor and 4-row cultivator. Power steering is standard. Note sturdiness of cultivator—stays in adjustment. Ask your dealer about low prices, easy terms.

You'll be hours and dollars ahead with a new Ford Tractor and Front Mounted Cultivator

JUST DRIVE IN TO ATTACH. Easy one-man job. No heavy lifting. Takes only a few minutes. To unhook, simply swing cultivator gangs forward, back tractor away. Quickly frees your tractor for other jobs.

For power farming at its best...

FORD

POWER STEERING IS STANDARD EQUIPMENT on all Ford row crop tractors . . . you don't pay a cent extra! Provides almost effortless steering, even in loose soil. You do better work, get more done.

UNIFORM PENETRATION is an outstanding advantage of all Ford front mounted cultivators. Parallel linkage holds pipe beams level at all times for uniform penetration . . . each gang operates independently for instant adjustment to ground contour.

ADJUSTMENTS THAT STAY PUT. Once you've made adjustments, they stay—there's no jumping on-and-off the tractor in the middle of the job. And Ford cultivators are fully adjustable to give just the setting you want. All adjustments are quickly and easily made, too.

CHOOSE THE TYPE YOU NEED. Two and four row units for wide and narrow row crops. Full range of attachments, including vegetable fertilizing and planting equipment. Choice of interchangeable front wheel arrangements on Ford row crop tractors. See them!

Now! a 3-plow diesel tractor at a gasoline tractor price*

"Experts" said it was impossible—but Ford has done it!

Yes, more than five years ago, a group of Ford's finest engineers set out to accomplish the "impossible." Their objective: to develop a powerful, fully equipped 3-plow diesel tractor to sell in the same price range as comparable gasoline tractors of other makes.* And they succeeded!

Save up to 50% or more on your fuel bills! So no longer must you pay a big premium to enjoy all the advantages of diesel power. Now you can cut your fuel

bills as much as 50 percent, and even more. Now you can profit from the extra lugging power in the new Ford diesels. Now you can enjoy all the convenience features for which Ford tractors have long been famous, *plus* diesel economy.

So stop in and see the new Ford diesel tractors at your nearby dealer's. Better yet, work-test one on your own farm, without obligation. Prove to yourself that you can do more, *save* more with a new Ford diesel than any other tractor in the 3-plow class!

Based on F.O.B. factory suggested list prices of other make tractors, as published when this advertisement was prepared.

-and only FORD has it!

Ford's Powermaster Series, introduced in 1958, included this three-plow diesel tractor. The company had reduced the cost of diesel engines so that both diesel and gasoline models sold at the same price.

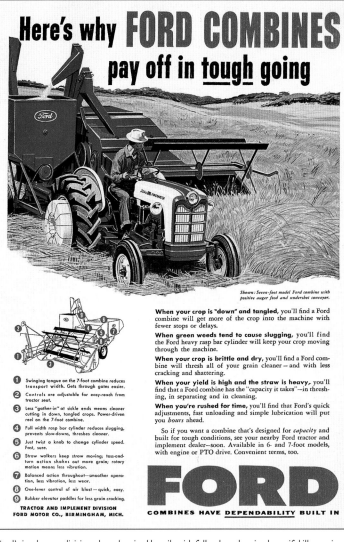

Here's why FORD COMBINES pay off in tough going

Shown: Seven-foot model Ford combine with positive auger feed and undershot conveyor.

When your crop is "down" and tangled, you'll find a Ford combine will get more of the crop into the machine with fewer stops or delays.

When green weeds tend to cause slugging, you'll find the Ford heavy rasp bar cylinder will keep your crop moving through the machine.

When your crop is brittle and dry, you'll find a Ford combine will thresh all of your grain cleaner—and with less cracking and shattering.

When your yield is high and the straw is heavy, you'll find that a Ford combine has the "capacity it takes"—in threshing, in separating and in cleaning.

When you're rushed for time, you'll find that Ford's quick adjustments, fast unloading and simple lubrication will put you *hours* ahead.

So if you want a combine that's designed for *capacity* and built for tough conditions, see your nearby Ford tractor and implement dealer—soon. Available in 6- and 7-foot models, with engine or PTO drive. Convenient terms, too.

1. Swinging tongue on the 7-foot combine reduces transport width. Gets through gates easier.
2. Controls are adjustable for easy-reach from tractor seat.
3. Less "gather-in" at sickle ends means cleaner cutting in down, tangled crops. Power-driven reel on the 7-foot combine.
4. Full width rasp bar cylinder reduces slugging, prevents slow-downs, threshes cleaner.
5. Just twist a knob to change cylinder speed. Fast, sure.
6. Straw walkers keep straw moving; toss-and-turn action shakes out more grain; rotary motion means less vibration.
7. Balanced action throughout—smoother operation, less vibration, less wear.
8. One-lever control of air blast—quick, easy.
9. Rubber elevator paddles for less grain cracking.

**TRACTOR AND IMPLEMENT DIVISION
FORD MOTOR CO., BIRMINGHAM, MICH.**

FORD

COMBINES HAVE DEPENDABILITY BUILT IN

Ford's implement division also advertised heavily with full-color ads using beautiful illustrations. This 1958 ad for the seven-foot-wide Ford combine could be driven by a PTO or by a combine-mounted engine.

CHAPTER SIX

INTERNATIONAL HARVESTER

The history of International Harvester Company (IHC) is a great tale of American business at work. The company was formed in a partnership between McCormick Harvesting Machine Company and the Deering Harvester Company in 1902. Both companies were already highly successful farm equipment manufacturers and had survived the explosive growth of the reaper business by the end of the 1890s.

Cyrus McCormick and William Deering had conversed 20 years earlier about combining their businesses to form a huge corporate farm machine company, but neither man could see eye to eye with the other. It took the next generation of each family to finally come to terms and allow a merger and consolidation.

Joining McCormick and Deering in the merger were Milwaukee Harvester, Champion Reaper, and Plano Harvester. Following the formation of the International Harvester Company, Keystone, Weber Wagon, and the Aultman Company were also absorbed.

However, it was not just a matter of company mergers. IHC had many good ideas and the company set about putting them into production, among them selling stationary engines which were to become international best-sellers.

IHC's first tractors were offered in 1906 and as business boomed the company built a new tractor manufacturing facility in Chicago in 1910. The Deering and McCormick companies retained their brand names and distribution systems retaining their existing dealers and adding IHC products to each dealer group.

IHC looked to the future and created a double line of large and small Titan and Mogul tractors. In 1915 it offered the Mogul Orchard 8-16 Tractor for $675.

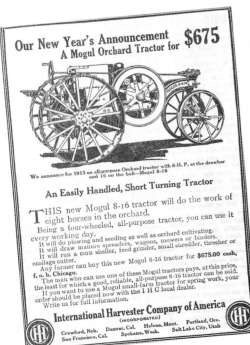

The International Harvester Company (IHC) was formed in 1902 and flourished as the parent company of both McCormick and Deering. IHC sold Mogul tractors through the McCormick dealerships. This 1917 ad for the Mogul 8-16 orchard tractor noted that it would do the work of eight horses in an orchard, that it handled easily, and that it was a short-turning tractor. Several variations of the 8-16 were built during this period.

IHC sold Titans through the Deering network of dealers. This Titan 10-20 was a very popular model at the end of the 1920s and helped IHC in their sales battle with Ford. By 1921, IHC had sold over $70 million-worth of them.

In the mid-1920s, the now famous "Farmall" tractor was introduced. It was the first effective general-purpose tractor. The Farmall evolved into a complete line, including the F-30, F-20, F-12, and F-14. In 1939, IHC revised the Farmall line and introduced the Letter Series, the A, B, H, and M, as modernized versions of the earlier Farmalls.

World War II brought with it a huge demand for mechanization. The manufacture of tractors, tanks, four- and six-wheel-drive trucks, and road construction equipment boomed at IHC. Their patriotic ads proclaimed "Today . . . a jungle. Tomorrow. . . a runway" and "Power and More Power for the Nation's Arms and the Nation's Farms." They built thousands of vehicles, and in the process won many Army and Navy "E" awards for their work.

In 1948 IHC produced its one-millionth Farmall, an M model. Widely proclaimed in their colorful print advertising, the company that "Nearly 9 out of 10 Farmalls built since 1923 are still on the job today." By this time the Farmall series had become the most popular farm tractor in America.

The Farmall Cub was, at first appearance, a "toy tractor," but it was as tough and hardworking as IHC's full-sized line-up. The Cub made itself popular for small row-crop farming and for mowing large acres of lawn.

The Super C was another popular model that could be ordered in row-crop, standard, and industrial variations, but it was especially suited to row-crop farming with its two-row, two-plow configuration.

Throughout the 1950s, IHC revised and improved an ever-widening model line-up including the 100, 200, 300, 400, and 600 series and all their variations. A change in the color scheme and styling took place in the late 1950s, and IHC went from all red to red and white. The 1960s and 1970s were difficult years for IHC; the company became embroiled in protracted UAW strikes and a tight money situation.

Since its earliest years IHC had also built trucks. This was a separate entity that had evolved from the building of powered-buggies to light trucks and eventually onto a full line of IHC road freight haulers.

The business was purchased by the giant Tenneco Company and merged with its other tractor holding, J. I. Case. A new company emerged: Case-IH, which today is Tenneco's largest division, and continues to build tractors.

In 1922, IHC offered the small 8-16 model equipped with a free P&O two-furrow plow for $670 f.o.b. Chicago. This Mogul model came with a belt pulley, fenders, a platform, a throttle governor, adjustable drawbar, angle lugs, and brakes.

The price war at the beginning of the 1920s had IHC slashing its prices dramatically. In 1922, the company offered the Titan 10-20 with a free P&O three-furrow plow for $700 f.o.b. Chicago.

This great line art ad from a 1923 McCormick-Deering highlighted the three-plow tractor with its 29-point ball and roller bearings. Its offset seat and steering improved operator visibility.

McCORMICK-DEERING No. 11 Harvester-Thresher. Built in two sizes, 12-ft. and 16-ft. cut. Sixteen-foot machine shown here.

McCormick-Deering also sold a line of harvester-thresher implements. This No. 11 was built in two sizes, 12 and 16 foot.

The McCormick-Deering No. 11 Harvester-Thresher

you buy TODAY will bring you every new 1928 feature!

FOLLOWING close on the heels of the company's most successful harvester-thresher year, International Harvester has announced a 1928 model which features a score of improvements and refinements. No radical changes—for none was necessary. Every improvement is the result of a sincere effort to make combine harvesting as simple, satisfactory, and efficient as possible.

The change that is most easily recognized is the new position of the engine and radiator assembly on the bull wheel side of the machine. The weight is more centrally located, lightening the load on the grain wheel. Also, the engine in the new position is closer to the operator and is easily accessible.

During the harvest season just past, hundreds of grain growers were unable to secure McCormick-Deering Harvester-Threshers, due to the great demand for combines of this make. The combination of this stored-up demand and the rapidly advancing reputation of these machines bids fair to swamp us with orders for the 1928 season. To avoid disappointing customers, our dealers are already taking orders for harvest-time delivery.

If you will telephone the local McCormick-Deering dealer, or come to our store, he will see that you get one of our new catalogs showing the McCormick-Deering No. 11 as it is now being built for the 1928 harvest season. Study the catalog and talk with McCormick-Deering owners; then you will understand why the McCormick-Deering is the outstanding harvester-thresher success.

INTERNATIONAL HARVESTER COMPANY
of America
[Incorporated]

201 Potrero Ave.,
San Francisco, Calif.

734 Lawrence St.,
Los Angeles, Calif.

McCORMICK-DEERING
FARMALL

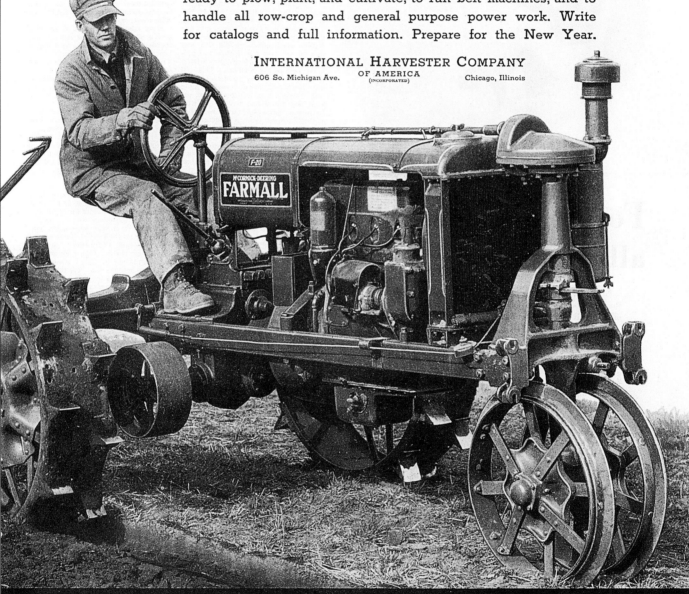

LOW-COST POWER the whole year 'round on drawbar, belt, and power take-off. Three FARMALL sizes: the 1-plow Farmall 12, the 2-plow Farmall 20, and the 3-plow Farmall 30 . . . each of them ready to plow, plant, and cultivate, to run belt machines, and to handle all row-crop and general purpose power work. Write for catalogs and full information. Prepare for the New Year.

INTERNATIONAL HARVESTER COMPANY
OF AMERICA
(INCORPORATED)
606 So. Michigan Ave. Chicago, Illinois

"IF IT ISN'T A McCORMICK-DEERING, IT ISN'T A FARMALL"

INTERNATIONAL HARVESTER
announces NEW 3-PLOW TRACTOR
the McCORMICK-DEERING Model W-30

Above: Miss Lillian Anderson, Queen of A Century of Progress, Chicago, at the wheel of the New Model W-30.

As the great tide of power farming surges ahead again International Harvester provides the 3-plow field with this new tractor, the McCormick-Deering Model W-30. The engineers are proud of this fine product, and every new owner will be.

The powerful W-30 is like a modern automobile in many particulars, but built for years of hard work in the field. Transmission and countershaft are *ball bearing*—the W-30 has 19 *ball* bearings altogether. And there are 14 *roller* bearings, generous use of which, in the front axle, steering gear, steering knuckle pivots, and elsewhere, makes for ease of operation. Triple seals in the front wheels and special diaphragm oil seals in the rear axle give perfect protection against grit and dirt. Replaceable cylinders and hardened exhaust-valve seat inserts add to tractor life.

The new W-30 is compact. Its turning radius is even shorter than in our own 2-plow McCormick-Deering 10-20! It handles as nimbly as an automobile. Available with steel wheels or low-pressure pneumatic tires. Ask the McCormick-Deering dealer about it.

THE PICTURES BELOW WERE SNAPPED AT KING FARMS CO.
This interesting farm enterprise at Morrisville, Pa., uses a large stock of McCormick-Deering equipment in its 2400-acre operations. These scenes show both regular McCormick-Deering and Farmall tractors with fast-working tillage outfits. Note the beautiful example of 4-row cultivation in the view below. Such Farmall work is a delight to the eye of every farmer interested in corn, cotton, or other row-crop farming.

You are now coming to the spring plowing, tillage, sowing, and planting operations of a more promising year. The McCormick-Deering dealer is *always at your service* for repair, replacement, and advice about new equipment.

THE NEW McCORMICK-DEERING O-12

The McCormick-Deering O-12 is a tractor that is especially designed to meet the needs of orchard and vineyard owners and open-field farmers requiring a light-weight, compact, low-priced tractor to pull a 16-in. or two 10-in. plow bottoms and other tools of proportionate size. The O-12 is equipped with big, soft, low-pressure tires for maximum traction in loose soil. Its transmission provides 2½ to 10¼ miles per hour speeds.

This same type of tractor is available also as the W-12, for general farming. The W-12 is usually equipped with steel wheels, but can also be had with low-pressure tires. Transmission range 2½ to 4¼ miles per hour. These small tractors will provide all the power needed on many farms, at a great saving in original cost and operating expense.

Photographs below from FORTUNE, *by Aikins, N.Y*

Below: These men make a great success directing International Trucks and McCormick-Deering Power Equipment at King Farms Co., Morrisville, Pa. Left to right: Karl C. King, Jack Cryer, A. C. Thompson, and Marvin Davis.

ACME *photo*

At Right: **CARL SCHOGER, OF PLAINFIELD, ILL.,** retained the championship last fall in the 56th Annual Wheatland Plowing Match, held near Plainfield, Ill.—and he did it with the McCormick-Deering Farmall and Little Genius Plow. Farmalls are now made in three sizes, 1, 2, and 3-plow. The dealer will demonstrate to suit your convenience. Get in touch with him early.

INTERNATIONAL HARVESTER COMPANY
606 So. Michigan Ave. OF AMERICA *(Incorporated)* Chicago, Illinois

The banner headline of "The First Day with a Farmall, A Red-Letter Day on any Farm" lets you know immediately that this new Farmall was something special. In 1935 IHC offered the models F-30, F-20, and the F-12. The nattily dressed operator of the unit illustrated looks as if he is racing his Farmall to the local dance.

The First Day with a Farmall
A Red-Letter Day on any Farm

LIFE'S best adventures come with the FIRST time in everything. Think of the boy and his first long pants, the joys and griefs that belong to the first day in school, the first dive in the old swimming hole, the first sweetheart . . . the first cry of the first-born when the boy reaches manhood.

We all remember what an adventure it was the day we first drove an automobile. Farmall farming is another real experience to add to all the other *firsts*. On many thousands of farms, where horses have *always* set the pace, the McCormick-Deering Farmall will take over the power burden this spring and improve the whole situation for years to come. Thrills don't last, but the all-around satisfaction in tractor power *will* last and grow—with every crop, every season, and every job that calls for power on the farm.

INTERNATIONAL HARVESTER COMPANY

606 So. Michigan Ave.   *of America* (Incorporated)   Chicago, Illinois

Now there are three McCormick-Deering Farmalls, the F-30, F-20, and the sturdy new F-12 in the picture below. Stop in at the McCormick-Deering dealer's store and try out Farmall power.

McCORMICK-DEERING

World War II brought a different direction in IHC's advertising. As they were building tractors "for the Nation's Arms and for the Nation's Farms" the ad illustrated, with a wonderful watercolor by Donald Milles, the work of war and the dreams of the future.

POWER and *MORE POWER!*

FOR THE NATION'S ARMS
AND THE NATION'S FARMS

When Uncle Sam calls for military tractor power *he gets action!* International TracTracTors, planned and engineered to the most exacting specifications, are *ready-made for the job.* The needed machines are rolling off the assembly lines, in factories long since equipped and tooled for production and operating at capacity.

For many years International TracTracTors have proved their power, economy, and solid worth in the service of Agriculture and Industry. That's why they are chosen now to add their mighty pull to the nation's growing defenses.

In the Army the familiar red exterior of these big crawlers is replaced with regulation olive-drab—*almost nothing changed but the paint!* Underneath you'll find the same great combination of relentless power and enduring stamina that makes International TracTracTors famous wherever hard work must be done...for the Nation's arms, and for the Nation's farms.

It will continue to be this Company's No. 1 job to supply all possible power for food and for defense — for the greater strength and security of the United States of America.

INTERNATIONAL HARVESTER COMPANY
180 North Michigan Avenue   Chicago, Illinois

INTERNATIONAL HARVESTER

SYMBOL OF SERVICE
TO INDUSTRY AND AGRICULTURE

INTERNATIONAL
HARVESTER

POST-WAR FARMING...
"ON THE CONTOUR"

WHEN WAR comes to an end, power-farming will move forward on pent-up plans. And the plans of every good-farming community will look to the saving of productive soil—to the control of erosion—to the sensible, modern practice of farming *on the contour*.

Again, in the advance of agriculture, Farmall and the Farmall System of Farming will lead the way.

Twenty-two years ago International Harvester introduced the Farmall Tractor... the first all-purpose tractor adaptable to all kinds of farming.

Today there are more Farmalls on American farms than all other makes of general-purpose tractors combined.

Farmall was FIRST... Farmall IS first today.

Farmall and the International Harvester Company are pledged to the faithful service of the progressive-minded farmers of the nation.

Tomorrow—as always—look to International Harvester for leadership in farm power and equipment.

INTERNATIONAL HARVESTER COMPANY, 180 N. Michigan Ave., Chicago 1, Ill.

INTERNATIONAL HARVESTER

In this full-color IHC ad, the ideals of peace and good farming practices were promoted using the Farmall line of tractors. The ad featured yet another watercolor illustration of Donald Milles.

TESTIMONY FROM THE TALL TIMBER

International Power takes the forest giants in tow and takes the job in stride

● Billions of feet of timber brought from the depths of virgin forests are eloquent testimony to the stamina and efficiency of International Power.

There, among the towering trees, International Crawler Tractors haul giant logs over the rugged terrain, chunking rocks and deadfalls out of the way with bulldozer blades that carve out their own trails to the landings.

For many years International Crawler Tractors have proved their power and their unbeatable economy. Their reputation rests on thousands upon thousands of jobs of construction and reconstruction for industry and agriculture.

Buy Victory Bonds and Keep Them

THERE ARE jobs by the thousands to be done—highways, airports, bridges, dams, housing, farming, flood control and railroad maintenance; jobs in forests, mines and oil fields; jobs in big cities, towns and villages. All of these jobs will be well done when they're done by dependable International Crawler Tractors, Wheel Tractors and Power Units.

INTERNATIONAL HARVESTER COMPANY
180 North Michigan Avenue   Chicago 1, Illinois

INTERNATIONAL HARVESTER
Power for Victory... Power for Peace

Good forestry practice, the purchase of victory bonds, and IHC crawler tractors were promoted in this beautiful watercolor ad, courtesy of testimony from the tall timber.

The Farmall C two-row tractor offered "fingertip farming" with Farmall's Touch-Control System which raised, lowered, and adjusted implements. The 1948 ad also noted that April is "Farmall C Month."

This full-color Farmall System ad illustrated the company's diversity in farm equipment including tractors, plows, and self-propelled and pull-along combines. It also noted that the farmer should stand by his Harvester dealer, as the dealer will stand by the farmer.

This great ad for Farmall tractors featured the headline, "Here's the millionth Farmall" and featured a bright red Model M—portrayed much like a Hollywood celebrity—as the center of attention. It also noted that 9 out of 10 Farmalls built since 1923 were still on the job.

Count on the FARMALL System

McCormick-Deering 123-SP Self-Propelled Combine opens up a field with no backswath. *Below:* The 123-SP, with pick-up attachment. Other McCormick-Deering combines — pull-type, hillside, rice.

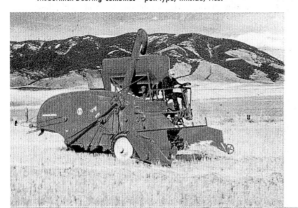

INTERNATIONAL HARVESTER, that is!

AMERICAN FARMING is *Power-and-Machine* farming. That is a big part of the story of American agriculture.

International Harvester is proud of the fact that its 113-year history is linked with the success of five generations of farmers in America. The big red combine of today is very different from the reaper that Cyrus McCormick trundled into the field in 1831. But it all began there. It began with an *idea*.

Harvester engineers have built thousands of *ideas* into machines since that beginning. The skilled workmen in our factories have multiplied those machines by the millions. And the American farmer has always known *how to put them to use!*

One of the greatest of those ideas produced the FARMALL—the original all-purpose tractor. Today, after 23 years, Farmall tractors and the machines and tools designed for Farmall operation have set a standard of farming success in every community.

Count on the FARMALL System to bring you great improvements in the year to come. Count on INTERNATIONAL HARVESTER. Many of you have been disappointed in the recent months, but it will be different soon. For this much is certain—we are all advancing toward a new era of better living for the family farm.

Your point of contact with the new equipment that is coming is your Harvester Dealer. Stand by him. He will stand by you.

INTERNATIONAL HARVESTER COMPANY
180 North Michigan Avenue   Chicago 1, Illinois

Tune in "Harvest of Stars" Sunday, 2 p.m.
Eastern Daylight Time. NBC Network

INTERNATIONAL HARVESTER

HARVESTER *Leads the Way* in POWER-and-MACHINE FARMING

The Farmall Cub in this ad must have made any dad proud. Donald Milles once again provided a stunning ad for the Farmall Cub which was a "bear for work." Introduced in 1947, the Cub remained in the line-up until 1964.

A Cub in Size...
but a **BEAR** for work!

and Now—
INTERNATIONAL HARVESTER PRESENTS
THE *Farmall Cub!*

- For all operations on farms of 40 crop acres or less—and truck gardens.
- For special operations on truck farms.
- For large farms that need an extra tractor.

That's the Farmall Cub, the first tractor in history that's *built right* and *priced right* for a vast new field of tractor owners.

The Cub is the newest member of the famous FARMALL FAMILY. It brings the advantages of the FARMALL* SYSTEM OF FARMING to the small, family farm.

It's a Cub in size, but "a BEAR for work." You get *big*-Farmall quality and design, plus scaled-down, small-tractor economy. And there is a full line of matched, quick-change, easy-to-control

implements designed especially for this tractor.

The smooth-running 4-cylinder 10 h.p. engine develops approximately 9¼ h.p. on the belt. It uses considerably less than a gallon of gasoline an hour. There's a comfortable, roomy seat ... ample crop clearance under the chassis ... and "Culti-Vision" to give you a clear, unobstructed view of your work.

Fit the Cub into your farming operations. See it as soon as we can get one to your International Harvester dealer. Get on the seat and drive it. You'll find it handles as easily as your car.

INTERNATIONAL HARVESTER COMPANY
180 North Michigan Avenue   Chicago 1, Illinois

*Observe National Farm Safety Week July 20-26.
Make every week Safety Week on the farm.*

*Registered trade-mark.
ONLY International Harvester builds FARMALL Tractors.

$545 f. o. b. factory
(Equipped as illustrated, slightly higher)
Attachments and implements extra

Hear James Melton on "Harvest of Stars" Every Sunday. NBC.

INTERNATIONAL HARVESTER

62

McCormick Super C was especially designed in 1954 with flexible front and rear track widths for row-crop work. This full-color ad noted that the Super C was the Cultivating Champ of its class.

Measure...Compare...Prove to yourself
the McCormick® Farmall® Super C
is the Cultivating Champ of its class

Here is the most practical way to buy a tractor:
Measure the tractor in relation to your particular jobs.
Compare it, feature for feature, with any other tractor.
Prove to yourself—by actual operation on your own farm—that the Super C is unmatched in the all-purpose 2-row, 2-plow class.

Measure your row-crop work. Compare the look-ahead view—instant response steering—effortless, 7-way implement control. Prove to yourself the Super C is unmatched for precise, all-speed, all-crop cultivating.

Measure all your work. Compare the Super C and your choice of implements for each job, for quality, for top efficiency. Prove to yourself unmatched Super C job range.

Measure operating ease. Compare the hydraulic seat, the double-disc brakes, the handy controls. Prove to yourself you ride in comfort, drive with unmatched ease on a Super C.

Measure pull-power by socking a plow down deep in hard ground. Compare the surge of power. Prove to yourself that pound for pound the Super C has unmatched pull-power.

Measure economy. Compare first cost, upkeep, trade-in. Prove to yourself unmatched economy—15 to 25 percent more work on a gallon of gas on job after job.

Measure years of use. Compare normal service life, performance and durability. Prove to yourself by talking to any user, that Farmalls pay for themselves over and over.

Try a Super C. Measure—compare—prove to yourself the unmatched pull-power and job range. Drive it yourself. Get in-the-seat proof, on your farm. See your IH dealer—prove to yourself that the Super C is Pull-Power Champ of its class.

INTERNATIONAL HARVESTER

International Harvester products pay for themselves in use—McCormick Farm Equipment and Farmall Tractors . . . Motor Trucks . . . Crawler Tractors and Power Units . . . Refrigerators and Freezers—General Office, Chicago 1, Illinois

Send for free Super C catalog—Get all the Facts!
International Harvester Company
P.O. Box 7333, Dept. SF-6, Chicago 80, Illinois

Gentlemen:
I want the whole story about Super C Pull-Power. Please send me the catalog.

My name is _____

My address _____

My IH dealer is _____

CHAPTER SEVEN

MINNEAPOLIS-MOLINE

The Minneapolis-Moline Power Equipment Company (MMPE) has built some of the most interesting, innovative, and stylish tractors of the past 90 years. Rooted in the Frost & Wood Company that was founded in 1839, and the Cockshutt Farm Equipment Company founded in 1877, the Minneapolis-Moline Power Equipment Company evolved in the 1930s as a brand entity of its own. The formation of the company follows the classic tale of takeovers and mergers which, as they progressed, resulted in the creation of a far

stronger and more diverse farm machinery company to which others in the marketplace looked for leadership.

Before Minneapolis-Moline Power Equipment Company emerged in the late 1920s, Moline Implement, one of the future partners in the company, absorbed the Abell Engine & Machinery Works, the Acme Steel and US Steel, Mandt Wagon, and Henny Buggy. This gave Moline Implement a strong base and a good source of steel with which to build many kinds of farm machinery.

Minneapolis Steel and Machinery Company grew by absorbing a mass of small machine manufacturing companies including Twin-City Iron Works and several seed drill makers. Twin-City tractors were a product of Twin-City Iron Works, which had been founded in 1889 as a heavy construction iron business in Minneapolis. They licensed a German stationary steam engine and began to make them for J. I. Case and Reeves & Company for use in their steam tractors. Later on, Twin-City Iron Works built complete tractors for the Bull Tractor Company.

By 1936, Minneapolis-Moline (M-M) had survived the worst of the Depression and continued to advertise extensively with two-color ads for its range of row crop tractors. This ad for the M-M Twin City Tractors offered its line of two-four row "Quick-on, Quick-off" Power-Lift Implements. Six models with optional rubber tires were offered in 1936.

In 1936 the Twin City brand name was still being used extensively but machinery with the M-M brand name was slowly taking over, especially the Universal J and M models. This ad features the range of implements and row crop tractors M-M offered in 1936.

Interestingly, the first tractors that Twin-City Iron Works offered for sale were built for them by an outside supplier, Joy-Wilson, which continued to manufacturer the Model 40 for Twin-City until 1920. Minneapolis Steel and Machinery Company was instrumental in many other minor mergers and buyouts.

The Minneapolis Threshing Machine Company also had a long and involved history of buyouts, mergers, and acquisitions. The roots of this company included the Advanced Thresher Company and Universal Tractor.

The Minneapolis-Moline Power Equipment Company was formed when three major tractor businesses came together due to the massive financial pressure of the impending Wall Street Crash of 1929. They were the Moline Implement Company, the Minneapolis Threshing Machine Company, and the Minneapolis Steel and Machinery Company. They merged only a short time before the stock market crash, taking much of the strain off each other by not being separate small businesses.

This merger brought with it the Twin City tractor line, which had been the pride of the Minneapolis Steel and Machinery Company—it was on these tractors that the new company built its reputation. From this point forward the business flourished, despite the troubles of the crash and the severe drought in the Midwest. A series of models based around the Minneapolis-Moline Universal Company was offered with varying engine outputs and layouts. But the revolution was yet to come.

One of Minneapolis-Moline Power Equipment Company's tractors that combined the most useful tractor traits arrived in 1938. The result was the UDLX "Comfortractor," and it was the first tractor with cabin bodywork. This stylish model featured full fenders, a cab heater, a radio, defrosters, and windshield wipers. However, in spite of all of these luxurious features and hard sales program, MMPE sold only 120 of these "plow the fields by day, drive to town at night" machines.

MMPE also began to specialize with models like the Kombination Tractor, followed by a series of tractors designated Model J, U, S, N, and E. These were crop-specific tractors for row-crops and orchards that could be equipped with a full range of Minneapolis-Moline implements. Its line-up expanded in the 1950s to include the Uni-Balor and other specialized planting and harvesting machines.

In 1969, Minneapolis-Moline merged with Oliver and Cockshutt to form the White Farm Equipment Company. The Minneapolis-Moline name was carried forward, but much of the new machinery was cross-branded with name tags on the same machines in later years.

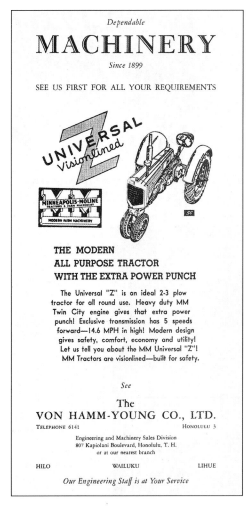

This post-war ad for the Minneapolis-Moline Universal Z was placed by Von Hamm-Young Company in Honolulu just after World War II. The Z was also available in standard tread width and was a very stylish row crop tractor for its time, with a top speed of 14.6 miles per hour.

This 1941 ad, entitled "This Land of Ours," courted farmers to better utilize their land with Minneapolis-Moline equipment. The Twin City brand name continued in use and nine different models were offered along with an extensive array of implements.

This colorful ad for the Visionlined series of M-M Universal tractors was creative and attractive. It featured a strange full-color process which was actually a separated printing process using four colors. The was one of earliest ads to use female tractor operators.

With the economy improving, along with printing techniques, Minneapolis-Moline's advertising became dramatically more colorful and refined, as this 1938 ad for the new Standard U Tractor displays.

Interested in Saving the Top Soil

Beauty is said to be only "skin deep", but, be that as it may, we must remember that, *comparatively*, the life sustaining depth of our earth is hardly that deep.

The photo above shows EROSION in the background. This is gulley EROSION, and, if not stopped soon may destroy the entire field—just as this "enemy" has already destroyed thousands upon thousands of acres. Even level fields are subject to SHEET EROSION—wind and water remove sheet after sheet of the top soil until almost all the fertile productive land is gone. Proper cropping and correct use of Modern Farm Machinery can help farmers everywhere to maintain the crop producing top soil not only for their own use, but for all the future generations. In the photo above is a painting, by the artist, Glen Ranney, showing one way how, by using the type of machinery now on the farms, EROSION can be *stopped* and *prevented*. THINKING FARMERS and others everywhere ARE INTERESTED in preserving the producing power of the soil, because a successful Agriculture is the foundation stone of our very existence. *Besides this*, WEEDS must be controlled and the FERTILITY must be maintained. All this means more work, but farmers have learned to depend on Modern Machinery to lift much of this burden from their shoulders.

Progress in taking drudgery out of farming has advanced more in the last fifty years than in all the ages before—and Minneapolis-Moline, continues the parade of progress for the benefit of all mankind. The top soil is still the most powerful force for good in the world.

Remember, Modern MM Machinery is worth waiting for—see your MM dealer for facts.

MINNEAPOLIS-MOLINE POWER IMPLEMENT COMPANY
MINNEAPOLIS 1, MINNESOTA, U. S. A.

This ad from 1946 preached the benefits of "Saving the Top Soil." This family scene, with father painting and the M-M tractor plowing the fields, was to remind farmers that they shared the responsibility for maintaining their farms in picturesque condition.

The Z Visionlined tractors were very popular with row crop farmers because of their strength and reliability. In this 1952 advertisement, M-M offered Uni-matic power to control the implement systems and a 36/31 horsepower four-cylinder engine.

A—MM UNI-HARVESTOR, one of six Uni-Farmor machines. B—2-Plow MM Model BF Tractor. C—MM FORAGOR. D—MM LS-300 MANURE SPREADER. E—MM 3-Plow Model ZB Tractor, with MM Hi-Klearance Plow.

MM UNI-HUSKORS won 1st, 2nd and 3rd in 1954 International Mechanical Corn Picking Contest!

the kids can't wait to take over!

You know it by the way a boy's eyes light up when he takes the wheel of a new tractor like the Minneapolis-Moline Model BF. Then a man can *appreciate* what *power farming* means to the young folks! Modern MM farm tractors and power-matched machines open a whole new future of progress and plenty for the boys and girls who stay on the land to become tomorrow's farmers. For power farming means *greater* production per hour, per acre . . . *lower-cost* production to *keep* the kids happy and prosperous on the farm.

Your Minneapolis-Moline Dealer is your community's sparkplug of power farming—the years-ahead MM way! See him and get the facts on the modern MM tractors and machines built to make farming easier and more profitable . . . an ever *better* way of life!

MM MODERN MACHINERY

MINNEAPOLIS-MOLINE
MINNEAPOLIS 1, MINNESOTA

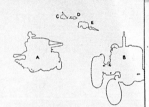

POWER-PACKED BF TRACTOR
Hitch this MM Model BF tractor to the biggest 2-plow job you can name. You'll discover power to spare . . . new fuel economy. New double-action Hitchor gives you hydraulic mounting of 3-point machines. Choose from a complete line of power-matched MM plows, planters and cultivators.

NEW 3-POINT TOOL BAR
Shown above with rigid shank cultivators, this new MM 3-point tool bar and carrier also mounts middlebreaker, spring tooth, lister planter, sub-soiler, and disc ridger attachments. Tool bar available in lengths from 60 to 184 inches.

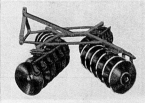

NEW 3-POINT DISC HARROW
Penetrate deep and clean in even hard, trashy soils with this new MM Disc Harrow. Frame is light, rigid, welded-steel. Gangs adjust independently from 15 to 25 degrees. Available with 20 or 24 discs for 6- or 7-foot cuts.

By 1957, Minneapolis-Moline had split up its advertising to concentrate on particular models and machines. With the right accessories, this Uni-Balor could be used as a windrower, a forager, a corn picker, a combine, or a hay baler.

NEW IN THE YEARS-AHEAD UNI-FARMOR FAMILY...

MM UNI-BALOR

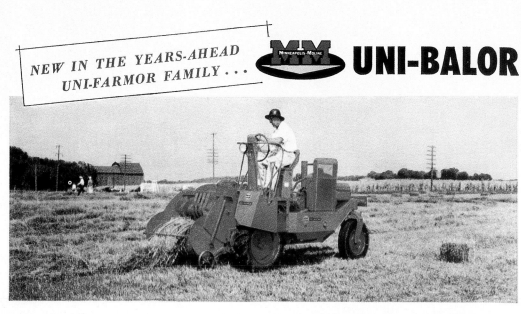

WITH ONE UNI-MACHINE OR A FLEET— MM UNI-FARMING PAYS YOU BETTER!

SEE THESE UNI-BALOR ADVANTAGES:

- Controlled auger feed
- Straight-through hay travel. No corners
- Exclusive twine tension control
- Big, 4-foot pickup gets all the hay
- Full size 14x18-inch twine-tied bales

Now, the same basic machine that gives you self-propelled windrowing, forage chopping, combining, corn picking, or corn picking-shelling, does your hay baling, too, to make MM Uni-Farming a still *better* way to farm! The all-new Uni-Balor for the champion Minneapolis-Moline Uni-Farmor takes this self-propelled system a big, new step ahead!

Uni-Farmor machines save you up to $1500 when you buy, save you money every time you add a Uni-Machine because every machine mounts on the same Uni-Tractor. And, when you Uni-Farm, you profit by all the champion advantages that have made the Uni-Farmor overwhelming winner in three straight major corn competitions.

Ask your MM Dealer to show you how Uni-Farming—with one Uni-Machine or a fleet—can pay you better than any other farming method.

UNI-WINDROWER—Open fields anywhere, cut hay or grain as crops ripen. Big, 10-foot capacity with hydraulic height control and constant cutting speed.

UNI-FORAGOR—Uni-Farmor capacity in the lowest-priced forage harvester on the market! Now with interchangeable direct-cut, row-crop or windrow pick-up headers.

UNI-HARVESTOR—Self-propelled operation *plus* famed MM Harvestor performance! Available with hydraulic-controlled 9-foot header or windrow pick-up.

NOW TOO
~~DISCOVER HOW YOU~~ CAN GO 100%
SELF-PROPELLED IN EVERY CROP AT
FAR LESS COST THAN PULL-BEHINDS!
and about half the price of other self-propelleds!

UNI-HUSKOR—The champion corn picker, in major competition and in your corn! Offers MM's exclusive husking bed with rotating and side-to-side action.

UNI-PICKER-SHELLER—Picks and shells corn with up to 30% moisture content in one trip through! You get the picking advantages of the Uni-Huskor and MM Shelling, too.

MINNEAPOLIS-MOLINE
MINNEAPOLIS 1, MINNESOTA

REMEMBER, ALL THESE UNI-MACHINES EASILY MOUNT ON *ONE* AND THE *SAME* UNI-TRACTOR—ASK FOR FACTS!

CHAPTER EIGHT
OLIVER

The Oliver Farm Company has its roots with an inventive Scotsman, James Oliver. Oliver invented a hardened "chilled" plow in the mid-1850s that resisted wear better than other plows because of the hardening process he put the steel through when building the plow.

Oliver formed his first company, the South Bend Iron Works, in 1986 in South Bend, Indiana, and went on to produce a long line of horse-drawn farm implements. Oliver later changed the company's name to the Oliver Chilled Plow Company and continued to build farm equipment for tractors as piston-

powered farming took over from the horse.

The late 1920s took a terrible toll on the farm machinery business. Oversupply, high interest rates, and the Wall Street crash contributed to a massive down-sizing of companies at the end of the decade. Over 70 percent of the companies that had existed in 1920 went out of business or merged.

The Oliver Farm Company developed in a similar fashion to the Minneapolis-Moline Power Equipment Company. It was formed in April 1929 by the merger of Oliver Chilled Plow, Hart-Parr, American Seeding, and

Nichols & Shepard. The new company had plenty of flexibility with a huge range of products to move ahead. The company later was renamed the Oliver Corporation.

Nichols & Shepard was also a farm machinery builder that started building Red River Special Threshers just after the turn of the century. The American Seeding Machine Company came about through a series of mergers of similar companies that brought with them an enormous production capacity.

One of the first new products was the Oliver Hart-Parr "Row Crop" tractor, a

Oliver's merger with Hart-Parr produced the new Oliver Hart-Parr "Row Crop" tractor in 1931. In profile it looked similar to the Farmall from McCormick-Deering with its tricycle layout and steel wheels. Oliver called these wheels "power on tip-toe." The tractor was good for plowing, planting, cultivating, and harvesting.

Oliver offered the "Plow Master" in 1944 with replaceable hardened steel alloy Raydex points on the tips of the plows. Oliver claimed they sliced through the soil more easily and gave longer life to the plow shares..

tricycle geared 18-27 horsepower tractor. It stayed in the line-up until 1937 as a two-model available in either three- or four-wheel configurations. The Oliver Standard 70 was introduced in 1935 and was kept in the line-up for many years.

A new streamlined model also emerged in 1935 and was re-streamlined as the Oliver Fleetline 60/70 series. After World War II these Fleetlines were upgraded with pneumatic tires to replace the steel wheels that had been part of every tractor since their earliest years.

The need for machinery that could handle heavier earth called for Oliver to introduce a line of crawlers. The two-plow OC3 appeared in 1953 and was offered as an "all-purpose, all-weather" tractor. Along with the HG model was offered a range of "Cletrac" models with track widths from 31 inches to 68 inches. Oliver expanded further in the early 1950s with a full range of farm machinery, from harvesters and bailers to threshers, combines, and other grain-handling equipment.

In the early 1950s, Oliver introduced a full line of diesel tractors with the 66, 77, and 88 models. The company also introduced the Super series for the 55, the 66, the 77, and the 88 models, available as either gas or diesel. The Super series' tough new stance was enhanced with the "Hydra-Lectric" hydraulic system and a three-point hitch as options. Many other small upgrades to transmissions, engines, and mechanical systems also improved the breed quality of the Oliver line-up.

In 1960, Oliver was acquired by the White Motor Corporation and two years later Cockshutt merged into Oliver's operations at White Motor Corporation. White went buying again in 1969 and purchased the Minneapolis-Moline Company. The group was then merged into the new White Farm Equipment Company.

However, the story of mergers and takeovers is not over yet! In 1980, the Texas Investment Corporation purchased White Farm Equipment and five years later Allied Products Corporation took over some of White's assets, adding them to its New Farm Equipment Corporation. Two years later they merged to form the White-New Idea Farm Equipment. White-New Idea Farm Equipment then represented the ideas, patents, and facilities produced from the merging and purchase of almost 50 farm machinery companies.

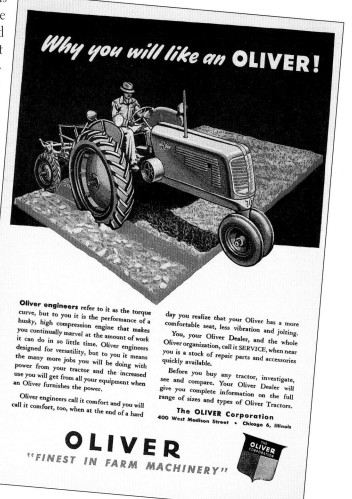

Why you will like an OLIVER!

Oliver engineers refer to it as the torque curve, but to you it is the performance of a husky, high compression engine that makes you continually marvel at the amount of work it can do in so little time. Oliver engineers designed for versatility, but to you it means the many more jobs you will be doing with power from your tractor and the increased use you will get from all your equipment when an Oliver furnishes the power.

Oliver engineers call it comfort and you will call it comfort, too, when at the end of a hard day you realize that your Oliver has a more comfortable seat, less vibration and jolting.

You, your Oliver Dealer, and the whole Oliver organization, call it SERVICE, when near you is a stock of repair parts and accessories quickly available.

Before you buy any tractor, investigate, see and compare. Your Oliver Dealer will give you complete information on the full range of sizes and types of Oliver Tractors.

The OLIVER Corporation
400 West Madison Street • Chicago 6, Illinois

OLIVER
"FINEST IN FARM MACHINERY"

One of the first postwar tractors from Oliver was the Row Crop 70. It came with standard rubber tires which made the tractor easier to handle and easier to drive. It also featured stylish streamlined bodywork.

In 1952 Oliver advertised three new diesel tractors. These were the 66, the 77, and the 88 models and all became very popular. These tractors had four- and six-cylinder engines, direct-drive power take-off, independent disc brakes, and "Hydra-electric" implement control.

"Teamed Up to Lick Tough Combining Conditions!" screamed this 1950 ad for the Oliver Cletrac Model HG crawler which is pulling an Oliver Model 15 Grain Master header.

The Oliver Model OC-3 crawler tells "Why it Pays to Visit Your Oliver Dealer." It adds "Extra Pull . . . for the tough acres!" Incidentally, very few farm equipment salesmen looked like the salesman in the ad.

Is There a Diesel Tractor for Every Farm Job?

Yes, Oliver and only Oliver makes all its models with Diesel Power

What are the sizes?

The Oliver "66" (two plow), the "77" (two-three plow), and the "88" (three-four plow). Row crop or standard models.

Will I save by buying a diesel?

Yes, but the amount depends on how much you use your tractor . . . the more hours you work it, the more an Oliver Diesel saves. It burns about 6 gallons of fuel where a gasoline engine burns 10 . . . cost-per-gallon is roughly half. Ask your Oliver dealer . . . he knows the figures for your locality.

Does it have as much power?

Yes. Oliver Diesels have the same horse-power rating as other Olivers in their class. They burn fuel slowly, start easily in cold weather, take heavy loads hour after hour.

Is the power as smooth?

With an Oliver, yes. Oliver gives you smooth, *six-cylinder* performance in the "77" and "88" . . . four cylinders in the "66".

Do I get all the Oliver features?

Every one! Features include: Six forward speeds, Grouped controls, Direct Drive Power Take-Off, "Hydra-lectric" implement control, easy-riding Oliver seat.

How can I tell if diesel is best for me?

See your Oliver dealer. He knows how fuel costs compare . . . knowing how much you work your tractor, he can tell you whether the fuel savings make the diesel your best buy. Remember, your Oliver dealer is *strictly impartial* . . . he sells tractors for diesel, gasoline and LP-gas, in all sizes. His only interest is to see that you get the very best power for your farm operation. See him soon! The Oliver Corporation, 400 West Madison Street, Chicago 6, Illinois.

The Oliver Corporation
400 W. Madison Street
Chicago 6, Illinois

F5-6

( ) Please send me literature on Oliver's complete line of Diesel tractors.

Name....................................

Address.................................

Town................RR.......State...........

OLIVER

THE OLIVER CORPORATION

"FINEST IN FARM MACHINERY"

The Oliver Super 55 was "The tractor that has everything." Available in diesel or gasoline with a 21-inch axle clearance. It was designed to take tool systems for farming or industrial equipment that could be pulled, pushed, or mounted at the front, side, or rear.

New! *TEAMED-POWER* does <u>both</u>
Husks and shells your 100-plus yields

Now you can husk part of your corn crop, shell the rest—*both* with Oliver's new husker-sheller. You invest in just one basic unit plus an economical husker, sheller, or both. That's Oliver's new TEAMED-POWER for modern corn farming—designed and built for the man who's stepping up to the 100- to 150-bushel class.

NEW GATHERING UNIT—saves your crop and saves you. It proves its mettle in capacity and easy handling—tops in safety, in shelled corn savings, and best at picking up the down corn. But now the story is even better. Oliver's new gathering unit shells less, wears longer, services easier.

THE HUSKER-SHELLER—two easy-change units, design-matched to suit your fondest production dreams. The husker: Oliver's famous valley design with eight 37-inch rolls, overhead fan, full-length corn-saver, and extra-wide wagon elevator. The sheller: the all-new, slow-speed, 1-section cylinder that practically eliminates cracking of the kernels.

Now you can be practical and modern with this new Oliver—TEAMED-POWER'ed for your 100-plus yields. Ask us for a demonstration on your own farm. "Pay-as-You-Produce" financing available.

THE OLIVER CORPORATION
400 W. Madison St., Chicago 6, Ill.

See Your *OLIVER DEALER* and <u>See</u>

This 1959 Oliver ad was bannered "Teamed-Powered" with a husker-sheller implement designed specifically for corn harvesting. The ad claims that this unit, teamed with the Oliver row crop tractor, could harvest 100 to 200 bushels per day.

the "Cut" that cuts costs...
OLIVER SELF-PROPELLED SP Combine

Oliver also produced a line of self-propelled implements including the Oliver SP Combine—it could carve out a 10-, 12-, or 14-foot swath in any standing grain crop. The machine featured six forward speeds, independent threshing speed, and live axle drive for uphill climbs.

It's the Oliver Self-Propelled 33, with *six forward ground speeds*—independent of threshing speed. With Vari-Draulic Drive (optional), you have constant threshing power at *any* ground speed.

Heavy stand? Think nothing of it! The big 33 wades right through, carving out a 10, 12, or 14-foot swath. Don't worry about weeds or undergrowth—the Oliver takes 'em in stride!

You run sure-footed over the sidehills ...low-mounted tank and engine make the 33 hug the ground. There's a *live axle drive* for uphill climbs—no chains to snap and cut you loose.

Your header has a 9-inch "give" to float over rough terrain (an especially valuable feature in the rice field model). The slats are always vertical—they can't shatter your grain.

And look over here! It's the Oliver 15 —six-foot, hundred-crop harvester—long a favorite for the average size farm. *See* these combines and how they save...see your Oliver Dealer!

OLIVER "FINEST IN FARM MACHINERY"

See your
OLIVER DEALER
and SAVE !

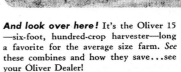

CHAPTER NINE

THE OTHER MANUFACTURERS — ADVANCE-RUMELY TO YUBA

Advance-Rumely; LaPorte, Indiana

The origin of this company dates back to 1882 when German immigrant Meinrad Rumely bought out his brother John's interest in a thresher-building business and established the firm of M. Rumely Company. To accommodate the fast growing agricultural economy of the Midwest, Rumely put ingenuity to work and produced one of the first straw-burning steam engines.

Meinrad Rumely died in 1904 but the business stayed in the family and contin-

ued growing for three generations. Meinrad's grandson, Edward A. Rumely, searched for a way to run the tractor inexpensively on liquid fuel. A deal with designer John Secor in 1908 led to the development of the first OilPull tractor. By 1910, 100 OilPulls had been built at the LaPorte, Indiana, factory.

The success of the business led to the purchase of Gaar, Scott & Company of Richmond, Indiana, in 1911, followed by the purchase of Advance Thresher Company of Battle Creek, Michigan. The following year North-

west Thresher Company of Stillwater, Minnesota, was also purchased.

Advance-Rumely maintained its reputation as a company that promised quality tractors. Their products were powerful and capable of handling any weather conditions. Owners of Advance-Rumely Tractors boasted that the OilPull's radiator never overheated. Another advantage of the tractors was the Triple Heat Control. This permitted the burning of kerosene, stove fuels, or distillate—successfully saving 30 to 50 percent of the fuel bill.

Advance-Rumely
This 1919 ad for the new 12-20 Rumely OilPull Tractor was created as line art for newspapers and magazines. It noted, "The Sweetness of Low Price never equals the Bitterness of Poor Quality." One of the great advantages of these OilPull Tractors was that the oil-filled cooling system would not freeze.

Advance-Rumely
"—and yet that OilPull never overheats!" This was a great ad line for these OilPull Tractors. The toughness of these OilPulls was renowned: Advance-Rumely claimed they would not freeze at even 40 below or overheat at 120-degree temperatures.

Advance-Rumely

Another great headline from Advance-Rumely was "Easy to Start and Always on the Job." They also claimed that the OilPull was, "The Cheapest Farm Power." This was backed by thousands of letters proving it from farmers.

Advance-Rumely's small OilPull was a 12-20 hp model. It had a two-cylinder, 6 x 8-inch engine and ran at 560 rpm. It had two forward speeds and one reverse and developed 15.02 drawbar horsepower with a maximum pull of 2,780 pounds. This model was in production from 1919 to 1924 and was perfect for a small farm.

Another notable OilPull was the Model L, a lightweight tractor with 15-25 hp. Its unit construction consisted of two pressed steel side members bolted to the transmission case with heavy studs which was then hot riveted to the front frame. The 15-25 used a two-cylinder engine with an operating speed of 730 rpm. Production of the Model L ended in 1927.

January 21, 1915, marked the end of the line for tractors with the Rumely name. The company went into receivership following a failed business deal in Canada. Under new management, however, the company continued successfully until June 1931. But a failed wheat crop caused heavy losses from a deal with Russia and as a consequence, Advance-Rumely was sold to the Allis-Chalmers Company.

Advance-Rumely had built 56,647 OilPulls in 14 different models by the time the company was sold. In addition, 3,192 Do-All tractors had been produced along with 40,000 grain separators and 26,000 combines.

Aultman & Taylor, *Mansfield, Ohio*

The Aultman & Taylor Company, formed in 1867, originally set out to manufacture and sell vibrating threshers. In 1892, the

Aultman & Taylor

This 1923 line art ad for Aultman & Taylor Tractors and Threshers was for traditional traction-style tractors and huge threshing machines sold as New Century Threshers. The ad features the 18-36 tractor, which was first built in 1915.

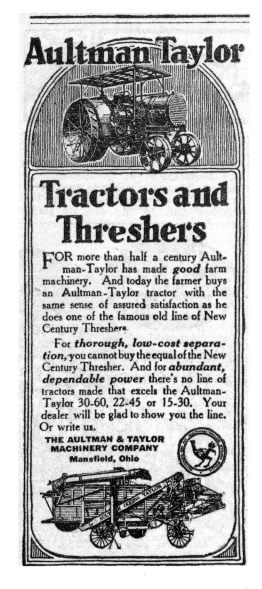

Avery

This huge double page ad from October 1907 featured some of the earliest use of photography in steam tractor advertising. It shows Avery Tractors being used for farming, road building, house moving, earth moving, and threshing. The tractor was sold with an engine but without a boiler.

company was reorganized into the Aultman & Taylor Machinery Company to sell and manufacture steam engines, threshers, and farm implements. Aultman & Taylor entered the light tractor market in 1918 with its kerosene-powered 15-30 Model. This model featured a gear drive, an engine cover, and a half canopy for driver protection. Some of this company's steam-powered tractors were gigantic and hundreds were sold, especially in the Midwest. In 1924, Aultman & Taylor was bought by Advance-Rumely and operated as a separate division until all the remaining A&T machinery had been sold.

Avery; Peoria, Illinois

Some amazing innovations have come about from inactive time spent in prison! Robert Avery developed some novel ideas about corn planters while imprisoned during the Civil War. Once released, he returned to Galesburg, Illinois, to form the Avery Planter Company in 1874 which specialized in planters, cultivators, and stalk cutters. By 1893 the company was building steam traction engines and threshers, and the company's vice president, John B. Bartholomew, was making improvements quickly.

Avery's Planter and Cultivator was built to handle two and four rows and its seven speeds allowed it to move slowly when necessary, yet it was able to cover 18 to 22 acres a day. It sold for $800 in the 1890s.

The Avery Undermounted Engine appeared in 1907, expanding Avery's tractor line. Promotions claimed, "You can move everything on top of the earth: haul logs, pull up stumps, move houses, haul ore, and pull just about anything on wheels or skids, etc., with an Avery Undermounted Engine." However, Avery had to work fast to enter the steam engine market and compete with Hart-Parr and other tractor companies. In 1909 the Avery Undermounted steam engine became the "farm and city" tractor marketed for use on small farms.

Avery

This neat Avery ad for the "Most Useful Machine on the Farm" offered a six-cylinder two-row version or a four-cylinder one-row version. Both could be equipped with three-speed transmissions and were "Sold at a Surprisingly Low Price."

After the deaths of the Avery brothers, Bartholomew became president. He purchased his first patent from Albert O. Espe, designer of the first Avery tractor and hired him full time. Design ideas were plentiful; Avery offered a full line of tractors, each only a slight horsepower difference from the other. Motor cultivators were part of their line and were marketed to the small farmers around 1919; however, they did not gain popularity soon enough against competitive manufacturers.

Advertisements claimed, "Satisfied Avery owners in every state in the Union and eighty foreign countries say you are not making a mistake when you get an Avery." To "Avery-ize" was to ensure bigger crops at a lower cost.

Avery's numerous designs led them to spread themselves too thin with parts inventories that were enormous. Avery continued to sell heavyweight tractors through the 1920s which was at a time when competitors were capitalizing on consumer demands for lighter tractors, not only for row crops but for general farming purposes. By this time, there were 186 companies producing 200 different models of

Bates

In October 1920, Bates Machine & Tractor Company from Joliet, Illinois, advertised "The Most Efficient Tractor in America." This half tract conventionally steered tractor weighed 4,900 lbs. and could generate sufficient power to operate an 18-inch Ensilage Cutter or a 28-inch Thresher with ease. Production ceased in 1937.

tractors in the United States. Avery found it could no longer compete and in early 1924 went into bankruptcy.

Bates; Joliet, Illinois

In 1919 a merger of the Bates Machine & Tractor Company and the Joliet Oil Tractor Company brought about improvements on the Bates Steel Mule that Joliet had been working on since 1915. Joliet's previous three-wheel design and high center of gravity had proved unstable on anything but flat ground. The Bates Steel Mule was successfully produced in five models by 1924.

The Bates Steel Mule Model D weighed only 4,900 pounds, had a four-cylinder motor with a bore and stroke of

NEW FEATURES
1919 BEAN TRACKPULL TRACTOR

Important improvements embodied in the 1919 "Bean" increase its efficiency by adding from two to ten times the wear-resisting strength of the parts affected. The 1919 Bean Track-PULL Tractor is the same design as the 1918 model, save for these valuable new features.

These improvements, added to the long proved design of the "Bean" make it a marvel of reliability and efficiency.

1919 VALUABLE IMPROVEMENTS

1. Motor—Improved oiling system. Improved carburetor control.
2. Track Rollers—Provided with hardened steel thrust bearings.
3. Drive Sprocket—Solid steel ring, machine cut, hardened and heat treated, assures long service.
4. Bearings—Provided with improved dust protected greasing facilities.
5. Rear Wheels—Provided with dust cap, wear resisting bushings and improved greasing facilities.

These, and many other features, are fully described in our new tractor catalog. Send for a copy before selecting your tractor.

See the 1919 model "Bean" at the Sacramento State Fair, August 31st to September 8th.

BEAN SPRAY PUMP CO. DEPT. K SAN JOSE, CAL.

Bean

This very strange looking 1919 Bean Trackpull Tractor was built in San Jose by the Bean Spray Pump Company. It featured a tricycle arrangement with a track drive up front, and two steerable wheels on the rear. The engine and transmission were hung to one side of the oval-shaped track. It weighed 3,100 pounds and, judging by the way the operator sat on this machine, it is a wonder he had any legs left after the harvest season.

The Belle City New Racine
The Thresher for the Fordson

With the Belle City New Racine Thresher ready to work you not only are sure of getting the grain threshed at just the right time but you can do a better job.

Important Belle City New Racine features that make this true are:

A Feeder governor for both volume and speed which positively prevents choking of the feeder and slugging of the separator.

Beater aids separation, spreads the straw across the straw racks, evenly distributed.

Four section straw rack gives ⅓ more agitation of the straw than most separators.

Long grate surface separates larger percentage of grain at the cylinder.

Ask your Ford dealer about the payment plan that makes it easy to save all your grain this year and thresh at just the right time.

BELLE CITY MANUFACTURING COMPANY, Racine, Wis.

Belle City

This 1926 ad for Belle City Manufacturing shows its Trackpull conversion of a Fordson Tractor along with one of its huge threshing machines designed to run with the Fordson. This crawler tractor conversion was one of the few cooperative projects with Henry Ford that produced decent results. The tractor operator looks like he's ready to take a lap round a racetrack rather than plow the lower 40.

4x6 inches, and featured an enclosed transmission and final drives. Its ads stressed that it was most efficient on belt work and did not pack the soil.

The glut of tractors on the market in the 1920s exacted a toll, in spite of the efforts of Foote Brothers Gear & Machine Company, which took over Bates for six years. Bates was forced to close its doors for good in 1937 along with almost 70 other manufacturers, whipped by the sluggish economy and lengthy testing procedures.

Bean; *San Jose, California*

The Bean Spray Pump Company from San Jose, California, was interested in row crop farming and promoted its new tractor, which had a daredevil top speed of 2 mph. The Bean 6-10 was produced from 1918 to 1920. It had wheels in the rear that could be steered and a track-type crawler in the front. The company revised the model as the 8-16 with a Bean-designed engine that featured a ball bearing crankshaft that was later adopted by International Harvester. The company also went out of business in the mid-1920s.

Belle City; *Racine, Wisconsin*

Belle City Manufacturing had a most unusual arrangement with Ford Motor Company. In 1926 they brought to market the Belle City "Trackpull" crawler conversion kit for the Fordson tractor. It offered "the ability to work under conditions where wheel traction is impossible." The kit was sold through Fordson tractor dealers and remained on the market through the beginning of the 1930s. Belle City Manufacturing also built a line of farm equipment, including the new Racine thresher which was heralded as "sure to get the largest percentage of grain threshed on time."

C. L. Best; *San Leandro, California*

In the 1880s on the Pacific Coast a great deal of experimenting was happening to make tractors less difficult to maneuver and less likely to mire down in the soft tule-type land.

Daniel Best got his start by developing a steamer with three wheels and a vertical boiler that was used extensively in the dense redwood forests of northern California to pull logs out of the woods. But on the soft peat soil of the Pacific Coast reclamation projects they were useless, a sell-out to Holt Manufacturing occurred in 1908.

In 1910 C. L. Best, son of Daniel Best, put together the C. L. Best Gas Traction Engine Company at Elmhurst, California. Three years later his first track-type crawler,

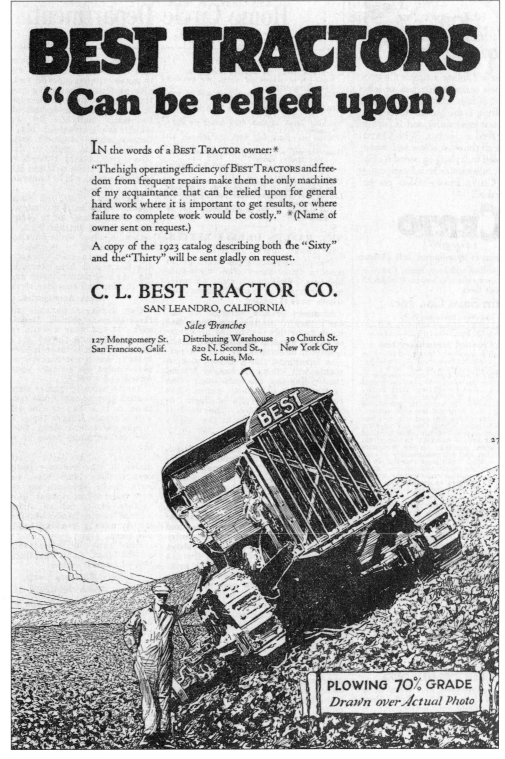

BEST TRACTORS
"Can be relied upon"

IN the words of a BEST TRACTOR owner: *

"The high operating efficiency of BEST TRACTORS and freedom from frequent repairs make them the only machines of my acquaintance that can be relied upon for general hard work where it is important to get results, or where failure to complete work would be costly." *(Name of owner sent on request.)

A copy of the 1923 catalog describing both the "Sixty" and the "Thirty" will be sent gladly on request.

C. L. BEST TRACTOR CO.
SAN LEANDRO, CALIFORNIA

Sales Branches

127 Montgomery St. Distributing Warehouse 30 Church St.
San Francisco, Calif. 820 N. Second St., New York City
St. Louis, Mo.

PLOWING 70% GRADE
Drawn over Actual Photo

C. L. Best

The Best Company was great at making claims for its tractors. Best touted endorsements from owners but failed to put the owner's name on the ad, saying that the name of the owner would be sent on request.

The Trundaar Tractor is a one-man outfit. Buy this tractor for service, and get what you pay for

When every day is worth $1000

Many days are worth more than that in busy seasons on the farm. How a Trundaar Tractor will save you many days a year

Massive Trundaar Transmission

Can you remember how many times in your business career you would have cheerfully given $1000 or more just to save a day?

Such occasions are only too frequent in the rush and drive of modern farm work, especially when labor is scarce and weather conditions are uncertain.

A Trundaar Tractor will save you many days in a year, besides enabling you to feel that, whatever the work may be, it is being done quickly and well.

You can work a Trundaar any day in the year, no matter what the weather may be like.

All the accumulated knowledge and experience of 23 years in the tractor industry has been built into the Trundaar Tractor.

Practical engineering principles develop ideal tractor

Its patented Trundaar Tread grips the ground so that the traction is positive, yet its area of 2160 square inches on the ground prevents soil-packing. If the ground is rough and uneven, traction is not impaired on the Trundaar, owing to the flexible nature and elastic operation of the double three-point suspension—another exclusive Trundaar idea. This suspension also absorbs all shocks and strains so that the power plant

and other parts of the tractor remain unaffected.

The Buckeye-Waukesha special tractor engine develops more than enough power for any possible requirement in farm work.

Trundaar owners have cut their fuel bills down to a minimum, due to the Buckeye-Deppé Integrator—exclusive equipment on the Trundaar. This system makes low-grade fuel operate like high-grade gasoline.

Dust cannot possibly affect the Trundaar as every vital part of this scientifically built tractor runs in oil.

A big, strong belt pulley is conveniently located for stationary work. It is always ready and is quickly connected.

Powerful multiple disc drive clutches control each tread—a new principle in tractor engineering.

The massive Trundaar transmission is built with a 20 per cent margin of safety so that it is practically wearproof. This transmission requires no differential and delivers maximum power to the drawbar.

Business farmers are writing us daily for our illustrated advance catalog, which describes the Trundaar in detail. Write today to our nearest distributor for your copy.

The Buckeye Manufacturing Co.
Anderson, Indiana

Trundaar Practical Belt Pulley

Trundaar Flexible Suspension

Buckeye-Deppé Integrator

Trundaar Tread and Driver

Trundaar Drive Clutch

Distributors for Northern California
HAMILTON & NICKELL
Sacramento

You can work a Trundaa. Tractor any day in the year

Trundaar Tractor
SPEEDS UP THE BUSINESS OF FARMING

Provides positive traction without packing the soil

Buckeye

The headline on the ad from the Buckeye Manufacturing says, "When every day is worth $1,000" which points out "How a Trundaar Tractor will save you many days a year." The ad claims that you can work a Trundaar any day in the year, no matter what the weather is like. The Trundaar 20-35 rolled on 2,160 square inches of track which helped to prevent soil-packing. It featured a new principle in tractor engineering—powerful multiple disc drive clutches to control each tread. This was the company's first ad in 1918, but by 1923 Trundaar had disappeared.

---

the Best "75," hit the market followed closely by two other models. Six years later Best introduced the "Muley," a track-type tractor that sold for $2,400, weighed 9,600 pounds, and had a four-cylinder engine. This model and its cousin, the 8-16 Pony, embody the basic principles of today's Caterpillar design.

In 1925, Holt and Best merged and the Caterpillar Tractor Company emerged. Holt was the first to use the term "caterpillar" and had it trademarked.

Buckeye; Anderson, Indiana

Buckeye Manufacturing entered the tractor business in 1912 with the Buckeye Junior, a half-tracked, tri-wheeled small tractor. Four years later Buckeye took over the Lambert Gas Engine Company which had built traction tractors since 1894. Lambert had also introduced the Steel Hoof, a lightweight three-wheeled cultivating tractor with self-cleaning steel wheels. In 1917, Buckeye introduced its first chain-treaded tractor and shortly thereafter, came out with its first Trundaar tractor, the 20-35, featuring a four-cylinder Waukesha engine and a tracked driving system.

The Buckeye-Waukesha engine featured a Deppé integrator induction system which raised fuel quality, while a flexible suspension system, massive transmission housing, special tread, and drive system all added to Buckeye's claim that the Trundaar "provides positive traction without packing the soil." By 1923, Buckeye and the Trundaar brand name had disappeared from the market.

Cascaden-Vaughan; Waterloo, Iowa

The city of Waterloo, Iowa, was the site of several tractor companies including the Waterloo Foundry Company, the Waterloo Gas Engine Company, the Waterloo Thresher Machine Company, and the Waterloo Motor Works. This is not to be confused with the Waterloo Machinery Company from Waterloo, Ontario, Canada, or the Waterloo Eagle Tractor from Appleton, Wisconsin!

In 1904 the Cascaden-Vaughan Company purchased the Waterloo Thresher Machine

WINNESHIEK

"IT SAVES THE GRAIN"

The Best is Always the Cheapest

Buy an up-to-date outfit. One that will please you and your customers by doing perfect work in any and all kinds of grain. Our patent oscillating device, the "missing link," does the business. It separates the grain from the straw.   :   :   :   :

Write for Our 1906 Catalogue

It will explain our threshing machinery in detail. Sent free on request.

CASCADEN-VAUGHAN COMPANY

Successors to the Waterloo Threshing Machine Co. and Waterloo Motor Works

WATERLOO   IOWA   U. S. A.

Mention The American Thresherman.

Cascaden-Vaughan
Cascaden-Vaughan was the successor to the Waterloo Threshing Company and the Waterloo Motor Works. This 1906 ad shows the Waterloo traction engine and its Winneshiek Thresher and claims, "It Saves the Grain." The pairing of these machines produced a powerful farming combination.

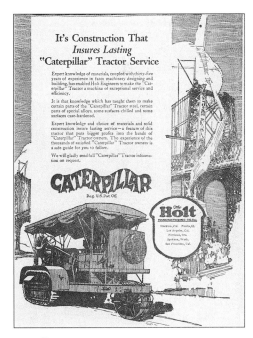

Caterpillar
In this 1917 ad, Caterpillar was still just a division of the Holt Manufacturing Company. Its tractors were designed for farming, construction, and engineering work. This ad features a Caterpillar 70-120.

Company and the Waterloo Motor Works. Its 1906 ad, showing the steam traction engine and the Winneshiek Thresher with "the missing link," did not offer much in detail.

Caterpillar; Peoria, Illinois

The history of the Caterpillar Tractor Company is a great epic of American business at

work. The bright ideas and talent of Benjamin Holt, and the merger of the Holt Manufacturing Company and the C. L. Best Gas Traction Company in 1925, brought about Caterpillar, a company that not only revolutionized modern farming but road, bridge, logging, railway, and airport construction as well.

The company's history goes back to a much earlier time, when Benjamin Holt, owner of the Stockton Wheel Company, built the first tracklayer-type steam-powered crawler tractor in 1890. Soon after, Holt changed Stockton Wheel to Holt Manufacturing Company.

Farmers in California's San Joaquin Valley were having trouble plowing the soft peat soil and Holt was approached to build a steam traction engine. The success of this huge tractor led to the building of more innovative steam-driven tractors using Holt's 1891 patent for the tracklayer drive system. The name "Caterpillar" was coined as a product name in 1904 for Holt's first crawler tractor.

In 1908 Holt purchased Best Tractors and the Colean Manufacturing Company in Peoria, Illinois, and began production of gasoline-

Drawn from photograph

Power, and Plenty of It

The "Caterpillar's"* field of usefulness is by no means limited to logging. There is a "Caterpillar"* of size and capacity for every power need. On farm or ranch, in the mining and oil industries, for building and maintaining roads, removing snow and doing other civic work—wherever tractive power and endurance are at a premium, the "Caterpillar"* has no real competitor.

In the depths of northern woods at 40 below zero, bringing out a ten-sled, 250-ton load of logs over a snow road demands the tremendous power and endurance which only the "Caterpillar"* Tractor affords. Old logging methods, necessarily slow and cumbersome, have given way to "Caterpillars"* because "Caterpillars"* put increased speed and capacity into every operation where the logging season of only two or three months makes continuous work essential for getting the logs out of the woods before the Spring break-up.

Up in the frozen woods of the North the "Caterpillar"* provides the same dependable performance as it does in bringing out mahogany logs in the Tropics, hauling teakwood in the Government forests of India, skidding logs in the Appalachian ranges and the big timber districts of the Pacific Northwest, and hauling in the hardwood swamps of the South. Experienced lumbermen and pulpwood manufacturers recognize the "Caterpillar"* as the one machine indispensable for swamping out and building roads, toting supplies to camps, and for transportation work of every kind.

Construction engineers, public officials, contractors, and industrial users everywhere, think first of the "Caterpillar"* when power, and plenty of it, is needed. Our illustrated booklet, "Caterpillar* Performance," will interest every power user. Copy sent on request.

* There is but one "Caterpillar"—Holt builds it. The name was originated by this Company, and is our exclusive trade-mark registered in the U. S. Patent Office and in practically every country of the world. Infringements will be prosecuted.

CATERPILLAR
Reg. U.S.   Pat. Off.

HOLT
PEORIA, , , ILL.
STOCKTON, CALIF.

THE HOLT MANUFACTURING COMPANY, *Inc.*
PEORIA, ILL.   STOCKTON, CALIF.
Export Division: 50 Church St., New York
Branches and service stations all over the world

Caterpillar

The T-35 was advertised as "The Supreme Small Tractor" and was offered at the amazingly low price of $375.

powered tractors. His Caterpillar crawler tractors became popular with farmers, construction companies, and the military. Holt's "Cats" were the models for British tank designs during World War I and his tractors were in high demand by the Allies for pulling supply wagons and artillery.

In 1925, Holt once again merged his interests with the Best family, this time with C. L. Best Tractors. This merger produced the Caterpillar Tractor Company. After the merger, combined sales were almost $14 million and gave the company a strong edge over its competitors. A couple of years later the purchase of the Russell Grader Company not only added to sales but broadened the product line with road graders and manufacturing facilities. Experiments on diesel engines further boosted sales as Caterpillar offered the first diesel-powered crawler tractor in 1931.

During World War II, Caterpillar built thousands of earthmovers, setting the groundwork for the most famous of the D-series, the mighty 320 horsepower D-9 bulldozer introduced in 1955. In the 1990s, the company continues to offer some of the finest tractors, earthmoving, and material-handling equipment built.

Centaur, Greenwich, Ohio

The Central Tractor Company was formed in 1921 and introduced its first product that same year, the Centaur Garden Tractor, a small cultivator-type tractor for row crops. In 1928 Central Tractor changed its name to the Centaur Tractor Corporation, continuing to build a line of garden, orchard, and truck farm tractors powered by LeRoi four-cylinder engines.

Their ads expressed the benefits of horseless farming, rugged construction, and labor conservation. These small orchard and vegetable farming tractors were popular for cultivation of row crops. Later in the 1930s, Centaur built a line of full-sized farm tractors that also featured LeRoi engines with a roller chain final-drive.

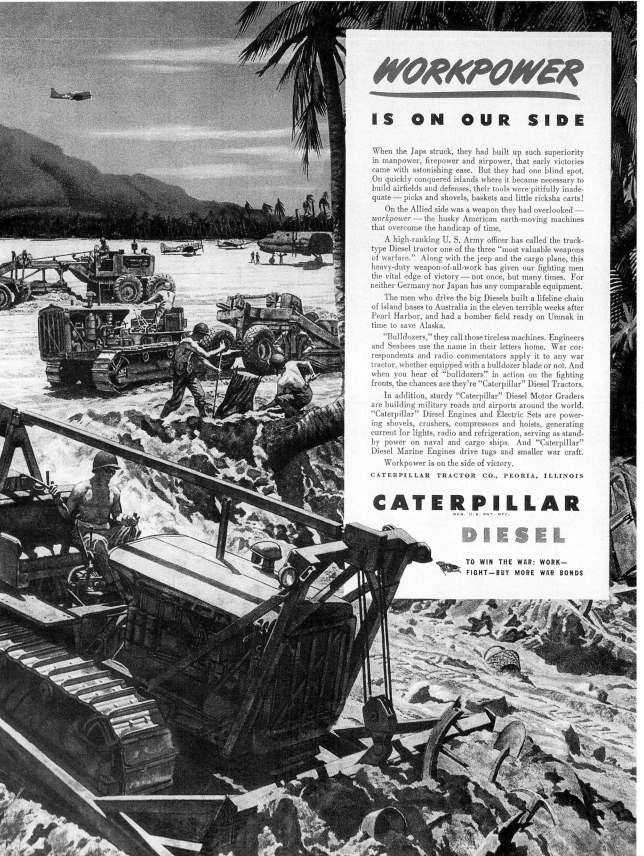

WORKPOWER

IS ON OUR SIDE

When the Japs struck, they had built up such superiority in manpower, firepower and airpower, that early victories came with astonishing ease. But they had one blind spot. On quickly conquered islands where it became necessary to build airfields and defenses, their tools were pitifully inadequate — picks and shovels, baskets and little ricksha carts!

On the Allied side was a weapon they had overlooked — *workpower* — the husky American earth-moving machines that overcome the handicap of time.

A high-ranking U. S. Army officer has called the track-type Diesel tractor one of the three "most valuable weapons of warfare." Along with the jeep and the cargo plane, this heavy-duty weapon-of-all-work has given our fighting men the vital edge of victory — not once, but many times. For neither Germany nor Japan has any comparable equipment.

The men who drive the big Diesels built a lifeline chain of island bases to Australia in the eleven terrible weeks after Pearl Harbor, and had a bomber field ready on Umnak in time to save Alaska.

"Bulldozers," they call those tireless machines. Engineers and Seabees use the name in their letters home. War correspondents and radio commentators apply it to any war tractor, whether equipped with a bulldozer blade or not. And when you hear of "bulldozers" in action on the fighting fronts, the chances are they're "Caterpillar" Diesel Tractors.

In addition, sturdy "Caterpillar" Diesel Motor Graders are building military roads and airports around the world. "Caterpillar" Diesel Engines and Electric Sets are powering shovels, crushers, compressors and hoists, generating current for lights, radio and refrigeration, serving as stand-by power on naval and cargo ships. And "Caterpillar" Diesel Marine Engines drive tugs and smaller war craft.

Workpower is on the side of victory.

CATERPILLAR TRACTOR CO., PEORIA, ILLINOIS

CATERPILLAR
REG. U.S. PAT. OFF.

DIESEL

TO WIN THE WAR: WORK—
FIGHT—BUY MORE WAR BONDS

Caterpillar

This World War II ad from 1944 shows a highly detailed color illustration of an air strip being built with Caterpillar tractors in the South Pacific. It is interesting to note that the bulldozer is pushing under an old wagon wheel, a shovel, and a pick.

Caterpillar

With the outbreak of World War II, Caterpillar was called upon to build an immense number of heavy-duty tractors and earth-moving machinery. This 1943 ad notes the Caterpillar had helped push the Alcan Highway through the wilderness at an amazing rate of eight miles a day.

AS FAST AS TROOPS CAN MARCH

How *fast can a road be built?* A mile in a week used to be called good in peacetime. The Alcan Highway was driven through the Northwest wilderness at the amazing rate of eight miles a day. But in North Africa the Army Engineers have smashed all previous records to smithereens. In the course of that campaign, they learned to fling roads across the desert at *four miles an hour*—as fast as men can march!

Leading the swift advance of these pioneer road teams are "Caterpillar" Diesel Tractors with bulldozers, filling gullies and stream beds. On their heels come more tractors or tanks, pulling heavy "V" drags that do the rough grading. And behind them, "Caterpillar" Diesel Motor Graders move steadily forward, finishing the job. The desert roads they build are not concrete speedways but they are serviceable routes for the long lines of supply trucks that bring up water, food, fuel and ammunition.

In every theater of war, "Caterpillar" Diesel Tractors, Graders, Engines and Electric Sets are making a sturdy contribution to Allied victory. Tough, powerful and dependable, they are able to run efficiently on almost any kind of fuel and in any climate. They haul guns, clear beachheads, slash through jungles, build airfields, generate current for lights and radio, and supply main or stand-by power for naval and cargo vessels.

With the great bulk of "Caterpillar" production going directly to the fighting fronts, the maintenance of thousands of older machines for vital work at home is increasingly important. That task is being capably handled by "Caterpillar" dealers everywhere. Their skill and specialized equipment are keeping "Caterpillar" Diesel power at work with a minimum expenditure of money and war-needed materials.

CATERPILLAR
REG. U.S. PAT. OFF.

DIESEL
Caterpillar Tractor Co., Peoria, Ill.

To Win the War: WORK—FIGHT—BUY U.S. WAR BONDS!

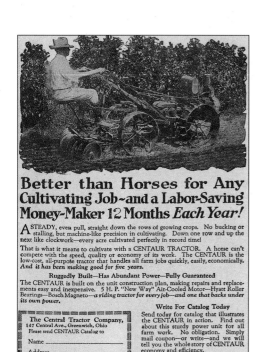

Central

The Central Tractor Company appeared in 1921 and offered the Centaur Garden Tractor. In 1928, its name was changed to Centaur Tractor Corporation. This Model 6-10 G sold for $515 and was designed for truck farms, orchards, and vineyards.

Cleveland; *Cleveland, Ohio*

O riginally known as the Cleveland Motor Plow Company, the Cleveland Tractor Company was formed in 1917 by a group of investors headed by Roland White. Cleveland Tractor Company's first diesel-powered crawler was the Cletrac 80 built in 1933. Many more models and track widths were manufactured over the years for light and heavy duty farming needs.

Like many tractor manufacturers of the time, Cleveland did not build its own engines but purchased different models from other companies, including Wisconsin and Hercules.

The Cleveland Tractor Company had some great ad lines, including "Geared to the Ground" and "Average Yearly Expense—$2.42." This second quote came from a letter written in 1927 by Archie M. Martin of Belgrade, Montana, who claimed that he had run his 1919 Cletrac for $2.42 a year (apart from gas and oil) for the last eight years.

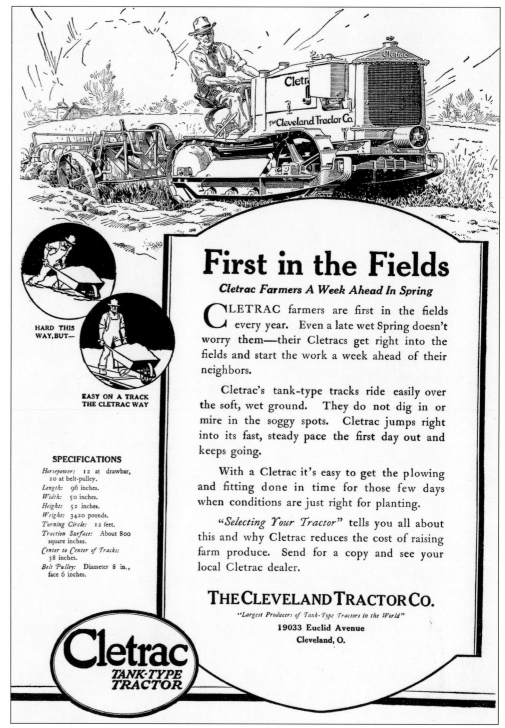

Cleveland

The Cleveland Tractor Company was the builder of Cletrac Tank-type tractor. Over the years, 40 different versions were offered from its Cleveland, Ohio, workshops. This 1921 advertisement called the Cletrac "First in the Fields."

A Great NEW 4-5 Plow Tractor

COCKSHUTT "50"

FEATURING
- The "thrifty 50" power plant. 273 cu. in. 6-cylinder engine.
- 6-forward speed transmission.
- Self-energizing disc brakes.
- "LIVE" Power Take-Off*.
- "LIVE" Hydraulic System.
- Choice of 4 Front Wheel Assemblies.
- Choice of Diesel or Gasoline Models.

*Cockshutt's famous "Live" Power Take-Off provides full, finger-tip controlled power for driven implements ALL THE TIME . . . as long as tractor engine is running.

■ Here it is! Cockshutt's newest, most economical 4-5 plow tractor. Power packed . . . power proven! The Cockshutt "50" has *all* the features that modern performance-wise farmers look for. It slugs its way through the toughest going with ease—yet its economy of operation will amaze you! The "50" is the latest addition to Cockshutt's complete line of tractors, ranging from the versatile 2-plow "20" (gasoline only) to the 2-3 plow "30", the 3-4 plow "40" and the 4-5 plow "50", all available in gasoline or diesel and with a choice of front end assemblies. Investigate now and see *why* it will pay you to make your next tractor a Cockshutt!

A COMPLETE LINE OF MODERN FARM EQUIPMENT

Tractors (Gasoline, Diesel, Distillate) • Tractor Mounted Equipment • Self-Propelled (Drive-O-Matic) Combines • Cultivators, Plows • Harrows • Manure Spreaders • Seeding and Fertilizing Machines • Forage Harvesters • Corn planting and picking machines • Feed Grinders • Wagons • etc.

COCKSHUTT FARM EQUIPMENT INC., BELLEVUE, OHIO. HEAD OFFICE—BRANTFORD, CANADA

For information on the complete Cockshutt line, write Cockshutt Farm Equipment, Inc. Bellevue, Ohio, or any distributor listed below.

Consumers Co-op. Assoc., P. O. Box 2359, Kansas City, Missouri.
Farm Bureau Co-op. Assoc., 245 North High Street, Columbus, Ohio.
Farm Bureau Services, Inc., 221-227 North Cedar Street, Lansing 4, Michigan.
Farmers Co-op. Exchange, Raleigh, North Carolina.
Farmers Union Central Exchange, P. O. Box G., Saint Paul, Minnesota.
Farmers Union State Exchange, 39th & Leavenworth Street, Omaha, Nebraska.
Indiana Farm Bureau Co-op. Assoc., 47 South Pennsylvania Street, Indianapolis 9, Indiana.
Midland Co-op. Wholesale. 739 Johnson Street, Minneapolis, Minnesota.
Pacific Supply Co-op., P. O. Box 1004, Walla Walla, Washington.
Pennsylvania Farm Bureau Co-op. Assoc. P. O. Box 23, Harrisburg, Pennsylvania.
Eureka Mower Company, Leonardsville, N. Y.
Green Harvester Company, Seaboard Park, Columbia, South Carolina.
B. Hayman Company Inc., P. O. Box 2847, 3301 Leonis Blvd., Los Angeles 54, California.
King Sales Company, 445 Tennessee Street, Memphis 3, Tennessee.
Kingston Implement Company, Roseville, Illinois.
J. W. Promenschenkel & Son, P. O. Box 255, Genoa, Ohio.
Southwest Company, P. O. Box 5647, 2811 Taylor Street, Dallas 2, Texas.

Keep ahead with COCKSHUTT

COCKSHUTT FARM EQUIPMENT — PARTNERS OF PROGRESSIVE FARMERS FOR 114 YEARS

Cockshutt

The new Cockshutt 50 ad from the late 1950s was drawn with an oversized tractor and an undersized farmer to emphasize the size of the tractor. Cockshutt was eventually purchased by Oliver in 1962.

Cleveland continued to offer crawler-type tractors throughout the life of the company, only differing from the basic design with a three-wheeled, rubber-tired, General GG model from 1939 to 1944. The Cleveland Tractor Company was famous for its 40 different "Cletrac" models, built until they became part of the Oliver Corporation in 1944. The Oliver Corporation was later purchased by the White Motors Corporation and the brand name "Cletrac" was discontinued.

Cockshutt; Brantford, Ontario and Bellevue, Ohio

The Cockshutt Plow Company was one of the earliest manufacturers of agricultural equipment. It was started in 1877 by James Cockshutt in Brantford, Ontario, to make horse-drawn farm equipment.

The company flourished and entered the tractor industry in 1924 by marketing other company's tractors, including Hart-Parr, Allis-Chalmers, and United Tractors with Cockshutt nametags. This early "badge-engineering" gave the Canadian company a new standing in the farm equipment business without having to actually design, test, or manufacture a machine of its own.

In the mid-1930s, the company took over the distribution of Oliver tractors in Canada and in the late 1940s opened a manufacturing facility to produce their Buda-engined Model 30. This popular model had an innovative power take-off design that utilized a separate clutch. By the early 1950s, Cockshutt was offering a line of gasoline, diesel and distillate (kerosene) tractors, combines, cultivators, forage harvesters, corn planters, and pickers, as well as a full line of tractor-mounted equipment and wagons.

The Cockshutt "50" was one of its most popular models in the early 1950s. This sleek and graceful tractor offered a plethora of useful innovations and accessories, including either a six-cylinder gas or diesel engine, a six-speed transmission, live-hydraulics, and disc brakes. The 500 Series, designed by the Raymond Loewy Studio, was introduced in 1957.

In 1962 the company was sold to the Oliver Corporation, with which it had

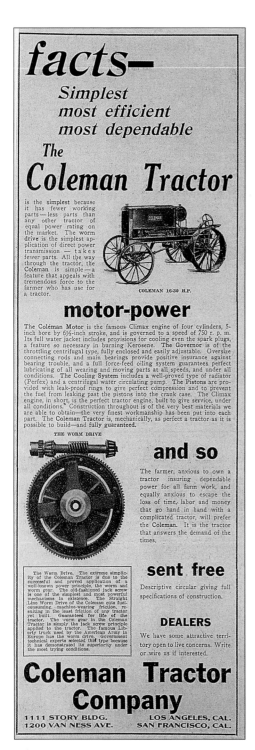

Coleman

The Coleman Tractor Company from Kansas City, Missouri, also had two West Coast offices. This 1920 ad featured the 16-30 tractor. The 16-30 was driven by a worm drive assembly that was fully guaranteed for the life of the tractor.

Common Sense

This ad by Herschell-Spillman Company from North Tonawanda, New York, is for the Spillman V-8 engines. The Common Sense Tractor Company used these engines in its strange tricycle-wheeled tractor. Common Sense was the first company to build a V-8-powered farm tractor.

had a long-term business relationship. Oliver was, in turn, absorbed by the White Motor Company in 1962.

Coleman; *Kansas City, Missouri*

The Coleman Tractor Company was an independent manufacturer. The company was in business for only two years and built only one single model. However, this single model was given two designations, the 10-20 and the 16-30. Powered by a four-cylinder, gasoline-powered L-headed Climax engine, the Coleman weighed close to 5,000 pounds. Sold as agriculture's "simplest, most efficient, most dependable" tractor, the Coleman featured a worm drive that was

Announcing A New Sandusky Tractor—"Model J"

$1095.00 F.O.B. SANDUSKY.

THE principal aim in developing this 10-20 tractor has been to widen the field of usefulness for Sandusky service. It solves the power problems confronting the farmer who does not require our large machine.

At our first conference, three years ago, covering such plans the argument was advanced: "Why not build a small cheap tractor which will sell itself on account of its low price?"

Our manufacturing facilities would enable us to do this on a large scale. But this did not accord with the policies of our President, Mr. J. J. Dauch, who personally controls seventeen large manufacturing establishments throughout the United States and Canada, each operated on the basis of "quality always."

This point having been determined, our first small tractor was designed and built. Farmers who saw this model wanted it; so it was with the second. From every indication these machines were far ahead of what was then being marketed. But they did not measure up to our ideal—a small tractor that we could afford to back with a binding guarantee; one that we knew would maintain the reputation thoroughly established by our larger machine.

The result is a tractor embodying our utmost efforts—not toward reaching a low price machine, but one to fill the need for a good tractor capable of real service at as low a figure as modern manufacturing facilities can produce and an efficient organization can market.

Engineers of world renown have approved its design, incorporating the famous Timken worm and gear drive, as representing the ultimate standardized small tractor. Years will prove that this Company has again set the pace of leadership in the Model J, as has been true in the case of our Model E.

10-20 Model J
Burns Kerosene

$1095.00 F.O.B. SANDUSKY.

TO his engineers Mr. Dauch said: "We are not interested in the manufacture of a cheap tractor. Our manufacturing and sales organizations are imbued with the quality idea embodied in our Model E. We cannot afford to demoralize these influences by putting out an inferior product merely to meet price competition."

These thoughts, together with the fundamental principle that this tractor must be adapted to the widest possible range of usefulness, dominated the development work in all its stages.

As a consequence the service which this tractor will render, the results it will produce in dollars and cents, the economies of labor which it will effect on your farm, make its purchase a profitable investment. The brief specifications below, and the more detailed description in our booklet, prove to any man that this machine offers a greater value per dollar than elsewhere obtainable.

Compare the Model J power plant with others. Consider its light weight—only 4,000 pounds; its Timken worm and gear drive, which actually increases in efficiency with use; its Hyatt Roller Bearing equipment. Remember that bevel gears and ordinary bearings mean friction, wear and loss of power. Remember our guarantees are based on the use of kerosene as fuel. Think! The result will draw your good judgment to this tractor like a magnet.

The Model J is a truly general purpose tractor, adapted not only to plowing, but also to the preparation of the seed bed, and all other farm power work, thus enabling you to profit by its service throughout the year.

It will be physically impossible to meet the tremendous demand of those who wait until their field work is in progress. We therefore urge you to write us at once, fully outlining your requirements. This will not obligate you in any way.

OUR GUARANTEE

We guarantee this machine capable of delivering a constant 2,000-pound drawbar pull. This is more than equivalent to handling two 14-inch plows at least eight inches deep in soil where a two-horse team can handle one 12-inch plow six inches deep, with a big reserve, or three 14-inch plows at depth formerly plowed with team and 12-inch bottoms. We also guarantee it to deliver 20 horse power on the pulley, which furnishes ample power for the operation of belt machinery up to a 24-inch separator with full equipment—all on kerosene as fuel.

We further guarantee this machine free from defects in either workmanship or material for one year from date of acceptance and will furnish f. o. b. factory any such parts returned, transportation charges prepaid, for inspection and credit.

TRY IT BEFORE YOU BUY

Continuing the policy created by us in the sale of our 15-35 Model E Sandusky Tractor, we will put the 10-20 Model J on your farm, demonstrating its ability to fulfill our guarantees under your own local conditions, and personally instruct you in the handling and care of this machine. It must prove itself profitably adapted to your requirements before you buy.

A NEW PLAN—FREE SERVICE

We announce an entirely new departure in tractor selling. A coupon book good for 100 hours of service, goes with every Model J. This means expert advice and help without charge when you want it. See catalog for details.

BOOK J-131—FREE

It describes in detail the principles of design, construction and ability of this smaller "Little Fellow With the Big Pull." It will enable you to make a thorough study of the real merits of this machine as compared with other tractors. In writing place your problems before us and we will gladly assist you in solving them without obligation on your part. Be sure to ask for Book J-131.

BRIEF SPECIFICATIONS

Our own specially developed four-cylinder, four-cycle, 4¼" bore, 5¼" stroke, quality motor. Drop forged high carbon steel crank-shaft, cam-shaft and connecting rods. Ninety-three percent tin base die cast interchangeable bearings. Valves enclosed. Combination force feed and splash oiling system. High tension magneto—impulse starter. Bennett air cleaner. A guaranteed kerosene burner. Surplus cooling capacity. Pulley for belt work. Adjustable cone clutch. Three speeds—2 and 3 m. p. h. forward and reverse. Timken David-Brown worm and gear drive. Hyatt equipped from motor to differential driving shaft, with universal couplings between motor, transmission and differential. This, together with three-point spring suspension, relieves road shocks, weaves and strains. All enclosed and protected against dust. Full steel construction. Short turning brakes. Weight, 4000 pounds. Length, 120 inches; wheel base, 76 inches; height, 52 inches. Rear wheel diameter, 48 inches; face, 12 inches and 6 inch extension rims; front wheels, 32 inches diameter, 6 inches face.

THE SANDUSKY TRACTOR
"THE LITTLE FELLOW WITH THE BIG PULL"

The Sandusky Tractors—both Models—are ready for inspection at and delivery from our principal Factory Branches and Service Stations at Indianapolis, Ind.; Bloomington, Ill.; Milwaukee, Wis.; Lewistown, Mont.; Minot, N. Dak.; Fargo, N. Dak.; Sioux City, Iowa; Lincoln, Nebr.; Wichita, Kans.; Dallas, Texas; Leesburg, Fla.; Los Angeles, Cal.; Sacramento, Cal.

THE DAUCH MANUFACTURING COMPANY, Sandusky, Ohio

DEALERS OPPORTUNITY

The 10-20 Model J Sandusky Tractor offers an exceptional opportunity to high grade dealers who will stock this machine on protected territory contracts. We want men of quality to represent us—men who are abreast of the times, in position to give service to our customers, and willing to give the time and energy required to build up a real business. For such men we have an attractive proposition.

Look for this announcement in the principal farm papers. It will be followed by the greatest big-space advertising campaign ever put back of a tractor. This in itself will create a big demand for you. Send for our dealers' prospectus. It outlines in detail the policies underlying this establishment; also our complete plans for such co-operation as will rapidly develop this account into a good-paying, permanent business for you with ever-increasing opportunities.

Write today—better still, wire territory reservation.

Dauch

This superb line art advertisement for the Model J Sandusky tractor is a stunning piece of line art. The ad ran in 1917 and the Dauch Manufacturing Company from Sandusky, Ohio, noted that its tractor was "The Little Fellow with the Big Pull." The company lasted only three years.

---

guaranteed for the life of the tractor. The Coleman Company ceased production in 1920 when it was absorbed by the Welborn Corporation.

Common Sense; *Minneapolis, Minnesota*

The Common Sense Gas Tractor was a startling idea for its time. The company started in 1915 with the first V-8-powered farm tractor. It was designed by H. W. Adams and featured a three-wheel layout which placed the third wheel at the rear. This huge driving wheel offered plenty of torque and little slippage with 40 horsepower at the PTO and 20 horsepower at the drawbar. The Common Sense used a Spillman "8" made by the Herschell-Spillman Company in North Tonawanda, New York. Herschell-Spillman supplied four-, six- and eight-cylinder engines to automobile, truck, and tractor manufacturers, and their V-8 was one of the earliest offered in this manner. Not only was the Common Sense tractor innovative from an engineering point of view, it was also snappy looking with a red frame and wheels, and yellow on the hood and rear fender.

The company used tractor schools as a method of teaching farmers about farm equipment and to sell their innovative tractor. In 1919 the Farm Sales Company took over the Common Sense Gas Tractor Company but failed to meet its sales requirements and ceased production in 1920.

Dauch; *Sandusky, Ohio*

J. J. Dauch, the owner of Dauch Manufacturing Company, entered the tractor business around 1910. Two years later Dauch offered the Sandusky 15-35 model with a strange boxy shaped coachwork covering a Dauch-built, kerosene-burning, four-cylinder engine. This model was revised in 1917 as the Model E with a new four-cylinder

Dill Special Tractor-Binder cutting heavy rice crop on Arkansas farm.

Section inside diamond shows Diamond Chain Drives used on this machine.

Proves Ideal Tractor Drive in All Farming Sections

In the soft rice fields of the South, as in the stiff clay sections of the North, tractors equipped with Diamond Chain Drive are demonstrating their greater economy, efficiency and endurance.

Farmers everywhere favor chain-driven tractors because chain drive not only delivers full power from engine to wheels, but is sturdy and flexible to withstand sudden strains and shocks. They know chain drive is more durable because its rolling action reduces friction and wear; also is easier to repair because it is simple and always accessible.

On thousands of farming implements, Diamond Chain Drive is giving highly successful service. Be sure any power machinery you buy has this dependable drive. Write us for free booklet—full of facts on drives for farm machines.

Diamond Chain & Mfg. Co.
Makers of high grade chains since 1890
INDIANAPOLIS, U. S. A.

DIAMOND CHAIN ◇ DRIVE
TRADE MARK

Diamond Chain

The Diamond Chain Drive accessory provided better power and traction delivery on all kinds of tractors. This 1920 ad shows the device in action on the Dill Special Tractor-Binder while harvesting rice in Arkansas.

engine. In 1917 the line was expanded with the Model J, a stream-lined, steel-wheeled tractor weighing close to 5,000 pounds. "We are not interested in building a *cheap* tractor" and "The Little Fellow With The Big Pull" were among some of the slogans from the Dauch Company, but even their most sincere efforts couldn't keep the Sandusky tractor on the market. It disappeared by 1921.

Diamond Chain; *Indianapolis, Indiana*

Diamond Chain was a manufacturer of chains for industry and agriculture and offered a special agricultural quality chain for chain-driven tractors. The chain drive allowed a simple way of adjusting the final drive ratio by changing the primary or secondary gear to obtain a different ratio. It also offered a "fix it in the field" operation because of its accessibility. Diamond Chain had special link shapes and was one of the best quality chains on the market in the 1910s and 1920s.

H. C. Dodge / Sprywheel; *Seattle, Washington*

Small walk-behind garden cultivators have been an ever-popular style of cultivator. Especially liked by truck farmers, seedmen, orchardists, and home gardeners, these light-weight cultivators could be manipulated in and about the crops with ease.

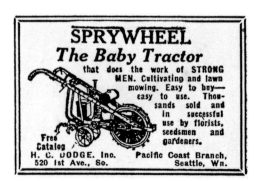

H. C. Dodge

In the years just after World War I there was an enormous number of small cultivator tractor manufacturers, especially on the West Coast. They created many variations of the walk-behind-powered cultivators including this Sprywheel Baby Tractor that the company claimed, "Does the Work of Strong Men."

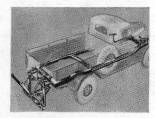

Dodge/Chrysler

In the post-World War II years, Detroit had excess production capacity for all kinds of automobiles, trucks, and tractors. The Chrysler Corporation saw the potential to sell many of its war-proven Power Wagon trucks to farmers for use in light-duty farming, saw-milling, and hauling.

Known as "The Baby Tractor," Sprywheel Garden Tractors were one of many variations built by manufacturers over the years. This early 1920s model featured an engine inside the front drive wheel.

Dodge; *Highland Park, Michigan*

During World War II, the Dodge division at Chrysler had turned out an enormous number of heavy-duty military trucks. In postwar years they had excess production capacity to build these war-proven, four-wheel drive Power Wagons. Like Willys, they were sold as an all-purpose vehicle for the farm. A line of power accessories was developed that could be driven off the power take-off which was already in the Power Wagon's drivetrain, or off the optional three-point hitch. They also developed a rear-mounted drill attachment, modified the tow hitch to pull a plow and harrows, included a winch, and presented the new four-wheel drive Power Wagon pickup at "one-fifth the cost of a tractor." The 1951 B-3 Power Wagon pickup was sold for $1,875 to $2,170.

Eagle; *Appleton, Wisconsin*

Eagle Manufacturing is one of the earliest of this century's tractor manufacturers. It was started in 1906, but only one model was built. The company reappeared in 1911 with a 56-horsepower four-cylinder engine. This was followed by two-cylinder models, including the 16-30 in 1916, the 12-20 in 1924, and the 20-35 in 1929.

All these machines featured a sideways-mounted radiator much like the early Hart-Parrs. Eagle advertising claimed their tractors were "the Simplest Tractor Built" and that "Simplicity is the key-note of the Eagle Twin-Cylinder Kerosene Tractor."

In 1930, the line was revised with a modern 6A model, powered by either a Hercules or a Waukesha four-cylinder engine. During World War II, Eagle suspended tractor production to focus on war materials and, when peace returned, tractor production was not resumed.

Simplest Tractor Built

What better guarantee of dependable, satisfactory service than Simplicity of Design? No delicate, complicated parts, needing frequent hair-line adjustments.

SIMPLICITY is the key-note of the Eagle Twin-Cylinder Kerosene Tractor. Very few working parts and every part works. Easy to understand; easy to operate. Every part easy to get at; therefore, easy to inspect and keep in perfect order.

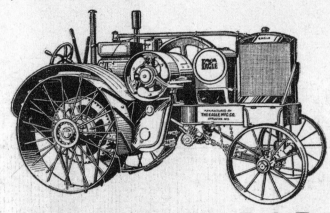

Remarkable Features of the Eagle Tractor

1.—Power Plant. A twin-cylinder, horizontal valve-in-head *slow speed* motor. The only kind that lasts under severe service.

2.—Simple, Powerful Clutch. Nothing could be simpler and easier to get at than this powerful friction clutch. Anyone can take it apart and put it together again quickly.

3.—Crank Case Door. A door fitted in the crank case which can be opened easily and through which you can easily reach the connecting rods and the crank shaft bearings for quick adjustment.

Could you think of a handier arrangement than this?

Always Ready for a Full Day's Work

Give the Eagle Tractor as careful attention as you do a good work horse and it will always be ready to serve you just as faithfully as your horse. It is Durable and Dependable.

Built in 3-plow and 4-plow sizes. Send us your name on a postal and receive our catalogue which tells all about the Eagle, the Simplest Tractor Built. You will like it.

EAGLE MANUFACTURING CO.
644 Winnebago Street   APPLETON, WISCONSIN

Eagle
The Eagle Manufacturing Company from Appleton, Wisconsin, started building tractors just after the turn of the century until the mid-1940s. In this 1920s ad for its Model H, the company claims it was the "Simplest Tractor Built." Once again, the ad is typical of the time in its use of fine line illustration.

Electric Wheel
The Electric Wheel Company began in 1890. By 1908 it had built its first commercial traction truck, and soon after, the first Model O tractor arrived. This 1919 ad notes, "Wonderful Power Built Into a Light Tractor." Electric Wheel also used product endorsement to attract future customers.

Electric Wheel; Quincy, Illinois

The Electric Wheel Company (EWC) was incorporated in 1890 with the purpose of manufacturing metal wheels by a new electrical welding process. Its roots go back to 1851 when it was the Brewster, Dodge & Huse Company which merged with Peru City Plow Company, the first builder of metal-spoked wheels in this country.

Tractor manufacturing experiments were made as early as 1904, and around 1912 the Electric Wheel Company presented its first tractor. During its years of production EWC produced a wide variety of machines from simple traction truck tractors to crawlers such as the huge 110 horsepower EWC 80. Production ceased in 1930 due to the strain of the Depression and was eventually absorbed as a division of Firestone Tire & Rubber.

One model of note from EWC was the Allwork 14-28 which was built from 1917 to

THE REEVES LINE

The Reeves Engines and Separators are the result of many years building experience and have proven reliable and efficient under the most trying conditions.

The Reeves Engine with its ability to start from any point without "Dead center" troubles — exclusive valve construction — patented drive wheel—strong rear axle extending across in rear of boiler, turning in very large boxes and not in the wheels, and many other points of superiority, is ideally adapted to the most trying service of all kinds.

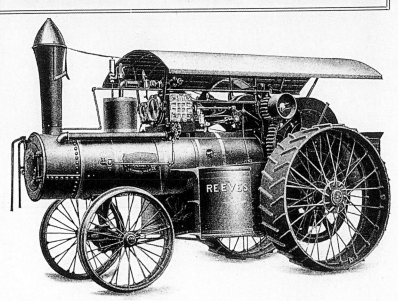

Emerson Brantingham Implement Co.,
  Regina, Sask.

Stalwark, Sask., March 20th., 1914

Dear Sirs:— In reply to your letter in regard to the Reeves Separator, it sure worked fine and we were well satisfied with it. It gets the grain all out of the straw and cleans it well. The elevator men claim that the grain from our machine was the cleanest that was brought in. The adjustable sieve is the best ever put into a separator and although we only have a 33 inch cylinder, we threshed 1,600 bu. of wheat a day with but six teams. It will handle all the grain that eight teams can bring to it, and all the farmers that we threshed for were well satisfied with the work. We could hardly tell what the separator could do on an average per day, as we seldom had teams enough to put it to its full capacity for a full day. Respectfully,

Signed — G. A. KOENING

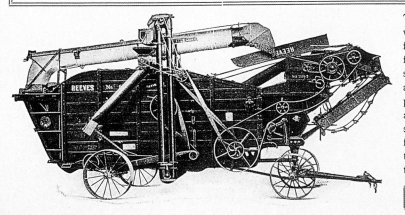

The Reeves Separator with its double separating features — extra strong frame — water proof construction—increased separating and cleaning capacity — improved cutter and feeder — special wind stacker and other superior features has reduced the thresherman's troubles to the minimum.

Write for Free Catalog

Emerson - Brantingham Implement Co.,
(Incorporated)

1169 W. Iron St.,

GOOD FARM MACHINERY

Rockford, Illinois

Ferguson

By 1948, Harry Ferguson had his own distribution system that sold Ferguson tractors and implements. This was the year of the infamous Ford versus Ferguson lawsuit (it was eventually settled in Harry Ferguson's favor). Most Ferguson advertising was done in black and white or two-color.

1924. It was powered by a four-cylinder engine with a bore and stroke of 5x6 inches. Unlike other tractor builders of the day, EWC used few vendor items in its tractor; the engine, clutch, transmission and final drives all were built by EWC. The 14-28 was a light, year-round, kerosene tractor used for pulling three plows and doing many kinds of field and belt work.

Emerson-Brantingham; *Rockford, Illinois*

The story of the Emerson-Brantingham Implement Company is another tale of a small company absorbing other smaller, businesses and, in turn, being absorbed by a larger enterprise.

John Manny and his two partners built reapers in 1852. They flourished, but after Manny's untimely death, the partners took over the young company, renaming it the Talcott-Emerson Company. Charles Brantingham joined them and within a short time became

Ferguson

Ferguson bannered his 1950 ad for the TO-20 tractor with the headline "World's Most Copied Tractor," a most perceptive view of the rest of the tractor industry.

Another Years-Ahead
TRACTOR BY
Ferguson
The Hi-40 with
Choice of 3
Front Wheel Styles

BRAND NEW

Ferguson

Ferguson was sold to Massey-Harris in 1953 although Massey Harris continued to offer the Ferguson separately. In 1956, Ferguson offered the new "Hi-40" which could be set up as a four-wheel, dual-wheel, or as a single-wheel tricycle tractor. That year Ferguson ran a free vacation contest called "Go places with Ferguson" that was open to everyone.

- QUADRAMATIC CONTROL
- "2-STAGE" CLUTCHING
- VARIABLE-DRIVE PTO
- DUAL-RANGE TRANSMISSION
- CONVERTIBLE FRONT WHEEL SYSTEM
- INCREASED POWER
- OPTIONAL POWER STEERING
- 12-VOLT ELECTRICAL SYSTEM

CHOICE OF MODELS. The Ferguson Hi-40 is available in models shown below. Front ends are also convertible—by the owner himself.

Four-Wheel Model	Dual-Wheel Tricycle	Single-Wheel Tricycle
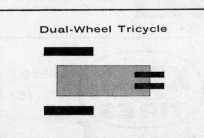		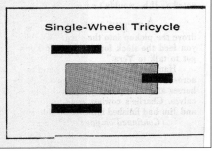

president. In 1910 Emerson and Brantingham re-formed the company, naming it the Emerson-Brantingham Implement Company. Within several years its line-up included not only implements but steam traction engines, and both kerosene and gasoline tractors.

Over the next 15 years the company purchased and absorbed a dozen smaller tractor and implement companies, including Rockford Engine, Reeves & Company, the Gas Traction Company, and the "Big 4" Tractor Works. Many different designs resulted from this merger, including the huge Emerson-Brantingham-Reeves 40-65 Kerosene tractor, built for Emerson-Brantingham by Twin City Tractors; a large three-wheeled Model L; a small four-wheeled 9-16 Farm Tractor; and a smaller three-wheeled Model 101 Motor Cultivator.

In 1928 the Emerson-Brantingham Implement Company was purchased by J. I. Case and merged into its general model line-up.

Ferguson; *Detroit, Michigan*

Ferguson was originally an English tractor manufacturer and entered into a manufacturing deal with Henry Ford in 1939 to build Ford tractors featuring the "Ferguson System." This process was the first integrated control system with built-in hydraulics, automatic implement protection, simple implement control, three-point conversion implement linkage, and a traction control system.

At the end of 1946 Harry Ferguson and Henry Ford terminated their agreement. After the split, Ferguson formed a new company, Harry Ferguson Inc., to manufacture and distribute his own tractors. The first American-built Ferguson tractor, Model TO 20, was built in Detroit, Michigan, at the Ferguson Park plant.

Powered by a four-cylinder Continental engine, it offered 16-22 horsepower. The line was quickly expanded with the more powerful "30" and the High "40" available in three front-wheel styles. Ferguson made an arrangement with Massey-Harris to form Massey-Harris-Ferguson in 1953. Soon

Vaughn

The "Flex-Tred" garden tractor is another of the early walk-behind orchard and row-crop cultivators built in the 1910s. Its claim to fame was a spiked, flexible tread traction system—it could rip your foot off if you didn't pay attention while driving it.

after, the Harris name was dropped and Massey-Ferguson went on to create an entire line of tractors.

Flex-Tred / Vaughn; *Portland, Oregon*

Small walk-behind tractors, such as the Flex-Tred garden tractor, are an important part of American tractor history. They not only gave the home gardener a small and useful cultivator but also gave orchardists and row crop farmers a useful tool. Powered by either a single- or two-cylinder engine, these tractors were mostly wheeled, but the Flex-Tred model from the Vaughn Motor Works offered a tracked drive system using four toothed wheels. A variety of attachments could be used with the Flex-Tred including plows, harrows, and discs, and in this 1917 ad Vaughn claimed that the Flex-Tred took "only 6 cents an hour to run."

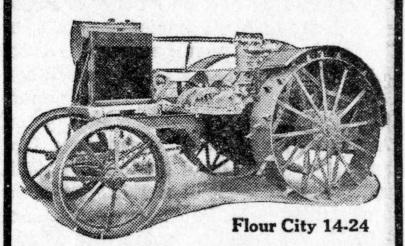
Flour City

Another tractor manufacturer in 1923 was Kinnard & Sons, manufacturer of the Flour City Tractor. The 14-24 pictured in this ad was "a powerful three-plow general farm tractor" that was sold at a very competitive price. The company closed in 1928 under the financial pressures of the coming depression and saturated tractor market.

Flour City; *Minneapolis, Minnesota*

The roots of Flour City Tractors go way back to the late 1880s. The company was started by partners O. B. Kinnard and Albert Haines, and initially produced hay presses. In 1896 the company offered its first traction engine and in 1900 its first production tractor. Kinnard & Sons Manufacturing succeeded Kinnard & Haines in 1917 and went on to build a wide variety of tractors over the next 30 years, including three-wheelers, high-wheelers, and stationery engines. The company eventually folded in the late 1920s with the decline of heavy tractors and the onset of the Depression.

Gladiator; *Los Angeles, California*

In the late 1910s, the Gladiator Tractor Company introduced, new to the market, the Gladiator Snake Crawler. It was touted as "the biggest small tractor and the smallest big tractor for the least money." It was designed for orchard and row crop farming and was powered by a Continental engine. It used a cast steel track, a roller pinion drive, and the body work allowed it "to crawl under trees like a snake." It was sold mostly in the Central Valley of California for orchard work and remained in business through the mid-1920s.

Gray; *Minneapolis, Minnesota*

The Gray Tractor Company had been manufacturing orchard tractors for some time before it introduced the 18-36 horsepower wide Drum Drive tractor in 1918. This unusual looking tractor featured a four-cylinder Waukesha engine. It was built with

101

Gladiator

The Gladiator Snake Crawler was a small cultivator tractor built in California for use in the Central Valley's vegetable and orchard farms. It was called the "Biggest Small Tractor and the Smallest Big Tractor for the least money." It must have been quite strong as it could produce 1,400 pounds drawbar pull.

simplicity and ruggedness in mind using a conventional steel frame and roller bearings where necessary. The 18-36 also featured a self-steering device, an aide to plowing, using a bar that rode in a previously plowed furrow to guide the tractor. The wide drum drive system was one of the Gray Tractor's greatest assets as the wide bearing surface distributed the weight evenly, did not pack the soil, and the spiked drum drive provided excellent trac-

tion. Gray claimed their tractors were "built for the man who wants good machinery."

Hart-Parr, *Charles City, Iowa*

The joining of Charles Hart and Charles Parr produced the Hart-Parr Company in Charles City, Iowa, in 1900. Advertisements claimed they were "Founders of the Tractor Industry." These engineering graduates from the

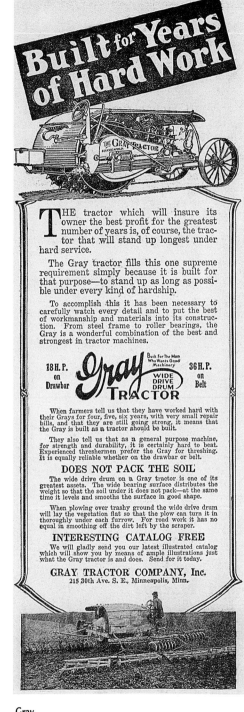

Gray

In its 1920 ad, the Gray Tractor Company noted that its Wide Drive Drum Tractor was "Built for the man who wants good machinery." This interesting variation of the tricycle tractor layout featured a very wide spiked rear wheel as the primary drive member. It was powered by a Waukesha four-cylinder motor.

Thorobreds

The $50,000 Hog from Mississippi

The Hart-Parr 30 from Iowa

"SCISSORS" the Grand Champion Duroc herd boar, didn't just happen to be a good hog. He is owned by Pinecrest Farm, Charleston, Mississippi, and was purchased from Mr. Ira Jackson of Ohio, who states that "Scissors" is the cumulative results of twenty-five years hog breeding experience. It takes years of study and careful mating to produce a grand champion hog.

The same principle holds good in building a tractor. Hart-Parr Company have been building tractors longer than anybody else, hence our tractor should be nearer perfection. Our Hart-Parr 30 is the logical result of twenty years experience in one line—it is a thorobred.

SPECIALIZING

Some tractor manufacturers refer to previous engineering experience in other lines as an indication of their skill in tractor building. It proves nothing. A good ration for hogs will not make a cow give more milk, and engineering principles that worked out well in an automobile or some other machine, will not necessarily work out in tractor building.

The Great Gran'daddy of All Tractors

Old Hart-Parr No. 1, built in 1901 worked successfully for ten years on a farm near Clear Lake, Iowa, and its last owner did not dismantle it until 1917. That's surely a wonderful record. Other Hart-Parrs have done as well and you may expect big things of the Hart-Parr 30 of today.

We produced the first successful oil tractor and have been building tractors continuously ever since. We are specialists. We have learned one thing and learned it well.

Our catalogue will be a source of valuable information for the tractor buyer. Copy mailed on request.

Long-Life Features of the Hart-Parr 30

One piece cast steel frame, making an engine bed solid as concrete—no bend, no twist.

A two-cylinder twin motor—fewer parts to wear out.

A slow-speed motor—750 revolutions per minute.

Force feed fresh oil lubrication, that keeps fresh oil on bearings at all times.

All working parts easily accessible, making it easy to keep the tractor in adjustment and repair. The Hart-Parr 30 is obviously simple.

The Hart-Parr Aftersale Service that teaches the farmer to take good care of the tractor.

Sturdy construction that dates back to the old Hart-Parr 60's with a record for long life.

Many of the old Hart-Parrs that plowed the virgin prairies of the Northwest are still in use today. The great grand-daddy of all Tractors was old Hart-Parr No. 1, built in 1901.

HART-PARR COMPANY
Founders of the Tractor Industry
451 Lawler St., Charles City, Iowa

A POWERFUL STURDY THREE-PLOW KEROSENE TRACTOR
HART-PARR 30
BUILT BY THE FOUNDERS OF TRACTOR INDUSTRY

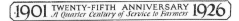

"Our Hart-Parr furnishes all the power for our 260 acre farm. We use it for plowing, discing, threshing, silo filling, shredding, and other belt work. We also do custom threshing amounting to more than 16,000 bushels a year." L. Schairer & Sons, Burlington, Illinois

HART-PARR
DURABILITY
insures long life and few repairs

Do you want a tractor that will serve you for about two years, or one that will work with high efficiency five or ten times as long as that? There are Hart-Parrs still operating after more than twenty years of farm work. Hart-Parr tractors are famous for their bulldog endurance. Every engine is made of big, strong parts; the crank shaft, main bearings, and connecting rod bearings are built for abuse—for brute force. Moreover, fresh oil is pumped to all working parts of the motor constantly—transmission gears run in a bath of oil. These are reasons why Hart-Parr tractors will do more work for a longer time than any in the market. And Hart-Parrs operate at a lower cost than other tractors, which is proved by hundreds of testimonials from Hart-Parr owners.

HART-PARR COMPANY   **CHARLES CITY IOWA**
FOUNDERS OF THE TRACTOR INDUSTRY

Get This Helpful Free Book on Power Farming

This big free book will convince you of the economy of power farming and also show you how to get the most for your money when you come to buy a tractor. Sign and mail coupon now.

FREE BOOK COUPON
HART-PARR CO.,
1027 Lawler St., Charles City, Iowa
Without obligation please send me your free illustrated book on power farming.

Name
Address

Hart-Parr

Hart-Parr's twenty-fifth anniversary was in 1926, and their advertising was headlined "Durability ensures long life and few repairs."

University of Wisconsin sold their first tractor in 1901 and spent the next two years developing a production tractor that they introduced as the 18-30 Model. Fifteen of the 18-30 were sold that year—an amazing number for a new tractor company. Designed mostly for heavy belt work, the tractor could also be used for light plowing.

Hart-Parr products were most recognizable because of their huge oil cooling radiators mounted like an outhouse at the front of the tractor. Using horizontally mounted two-cylinder engines they featured a long torque production stroke on all their engines. Strokes of 13 inches and above were not uncommon.

Their success with the 22-45 horsepower version brought another Charles into the picture, an investor and later partner, Charles D. Ellis. This joining of forces combined the financial support with the engineering brainpower necessary for the company to grow actively over the next 20 years.

They built an improved 17-30 model, which was put on the market in 1903, and followed it in 1907 with the now-famous 30-60

Hart-Parr

"A surgical operation not necessary." This comical ad from Hart-Parr for the Model 30 shows a surgeon with a saw ready to operate, but a friendly farmer waves that "I can get at any working part in less than five minutes." The comparison is rather chilling.

"Old Reliable" tractor. These machines were internal combustion-powered but built to look like steam traction engines with oil cooling, enormous flywheels run by a long-stroked, two-cylinder kerosene engine, which Hart-Parr advertising claimed could be run on any fuel, including alcohol. During the next 20 years Hart-Parr specialized in these large tractors for thresher men and heavy plowing.

In 1914 Hart-Parr introduced the "Little Red Devil," another strange three-wheeled tractor featuring a two-cycle, two-cylinder engine that obtained reverse by running the engine backwards. In its two years of production, between 3,000 and 4,000 were built. The Little Red Devil's strange layout, with a centrally mounted huge single rear wheel, caused it to be rather unstable on anything but flat ground.

Little Husky TRACTOR

Compact Powerful Durable

8=16 H. P.

Built to meet the severest tractor plowing conditions.

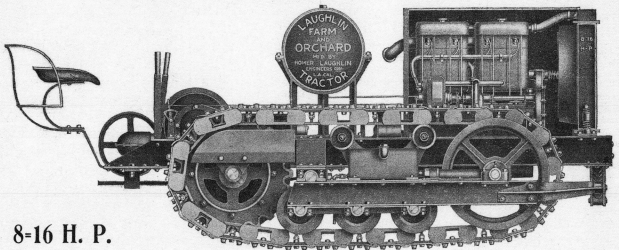

There are a number of patentable features on the **Laughlin "Little Husky"** Tractor which have been developed and perfected by a man who has followed the trend of the creeper type of tractor since its inception. He has studied and overcome the difficulties of the past and has produced a dependable machine, embodying improved features.

The oscillating frame is hinged independent of the sprocket axle and takes all the strain off from the track drive gears.

The track links are really a solid railroad tie with no loose spools or pins to work out and delay your plowing.

The "LITTLE HUSKY" Tractor is not an Experiment

BUILT BY——

Some Distinctive Features

Valve in Head Motor—Enclosed Valves.
Two Independent Side Clutches—Enclosed.
Large Cooling Fan—Gear Driven.
External Cut Drive Gears—Enclosed.
Spring Mounted Radiator—Perfex.
Stationary Drive Pulley.
Dixie High Tension Magneto—Impulse Starter.
Patented Sure-Grip Track Links.
Laughlin Heavy Duty Slow Speed Motor. 700 R. P. M.
Demountable Sprockets, Rim Mounted on Hyatt Bearings.
Double Flanged Idler. Mounted on Hyatt Bearings.
Transmission—Sliding Gear Type, Enclosed—running in oil.

Sent on request—descriptive literature, giving full specifications.

The Homer Laughlin Engineers Corporation

2652 Long Beach Ave. **Los Angeles, Cal.**

Homer Laughlin

The Laughlin Little Husky Tractor was offered as a farm and orchard tractor and was built in Los Angeles, California. Like the Gladiator, the Little Husky was built to handle farm work in California's Central Valley. The 8-16 horsepower model featured a Laughlin heavy-duty slow-speed motor with a 700 rpm maximum. It is interesting to note that the Little Husky Tractor "is not an experiment."

As Reliable as Horses and Much Faster !

YOU can depend upon your Huber Light Four to serve you as steadily and dependably as your horses ever have. It is ready to work, every day, all day, unaffected by heat, flies, or fatigue

The Huber Light Four pulls three plows and turns an acre an hour. It gets the plowing done on the few choice days when the ground is just right and assures a bigger yield. It works the ground more thoroughly. It saves time at harvest and does all belt work, even running a light thresher.

The Huber Clutch

Among the features that contribute to the exceptional dependability of the Huber is the counter balanced friction clutch. The friction arm is screwed on the end of the transmission shaft assuring perpetual rigidity. Pull the lever and the friction collar moves toward the motor expanding the friction shoes against the inside of the flywheel. There is no end thrust on the motor or transmission. The large friction area assures a tight grip and makes smooth engagement possible. This is only one of many Huber points of superiority.

The makers of the Huber Light Four are pioneer tractor builders with 20 years' tractor experience behind them. Every year they have won fast friends by making a reliable and dependable tractor. Huber tractors are firmly established. They are here to stay. And the makers will always be ready to give you the support you have a right to expect from the manufacturer.

You are sure of the service you will get from the Huber. Write for booklet "The Foundation of Tractor Dependability."

THE HUBER MFG. CO.

119 Center Street   Marion, Ohio

Canadian Branch—Brandon, Man.
Makers also of the Huber Jr. Thresher

12 H.P. on Draw Bar
25 H.P. on Belt Pulley

THE HUBER Light Four

Draws three bottoms
Turns an acre an hour

"THE TRACTOR DEPENDABLE"

Huber

The Huber Company had its roots in 1865 when Edward Huber invented a hay rake. Huber patented numerous agricultural implements, steam engines, and tractors. This 1920 ad shows two sad looking horses watching as a Huber Light Four takes over their jobs. The ad is bannered "Reliable as horses and much faster!" Large diameter front wheels were a Huber trait.

Hart-Parr's trend to downsize farm tractors continued with the introduction of the 15-30 Type A, the 18-35 Crop Maker, the 10-20 Model B, and the 20-40.

In 1918 the "New Hart-Parr" 12-25 tractors were introduced—these looked more like today's conventional farm tractor. With this new model the company added a pressure lubrication system, a new transmission, and a magneto ignition.

It was claimed in their advertising in 1921: "I can get to any working part in less than 5 minutes" and "A surgical operation not necessary." By their twenty-fifth anniversary in 1926, Hart-Parr was heralding their products for "Durability, long life and few repairs."

Late in the 1920s, both Charles Hart and Charles Parr withdrew from the business, the famous tractor wars, changing technology, and new trends proving difficult for them to accept. Charles Ellis, the third partner and original financial backer, bought Hart's and Parr's shares in the company and was left in control. In 1928, he turned over control of the company to his son, Melvin W. Ellis.

In 1929, Hart-Parr, the Oliver Chilled Plow Company, Nicholas & Shepherd, and the American Seeding Machine Company merged to form a single new entity—Oliver Farm Equipment Company. The Oliver Farm Company later re-formed into the Oliver Corporation, which was later purchased by the White Motor Corporation.

Homer Laughlin; Los Angeles, California

There were many small tractor manufacturers in California in the first 20 years of the twentieth century. Stockton, the San Francisco Bay area, and Los Angeles were centers of tractor building activity on the West Coast. These small

Intercontinental
This ad from Mexico for the Intercontinental C-26 three-plow tractor notes that horses have been retired and that Intercontinental Tractors are taking over.

companies focused initially on building tractors for rapidly growing vegetable and orchard farming in the Central Valley of California, which today is called the Salad Bowl of the World.

To meet the needs of this initially small-scaled but intensive farming, companies like the Homer Laughlin Engineers Corporation built small tracklayer-type tractors, suitable for row crops and orchard work. The 1919 "Little Husky" tractor offered 8-16 horsepower by a company-built four-cylinder gasoline engine. Interestingly, it was claimed that some of the design was "patentable" but not patented. With the seats mounted so far back and out in the open, the Little Husky must have been a strange device to operate. Homer Laughlin

offered several versions of the Little Husky including a 10-20 model but left the tractor business to others within a couple of years.

Huber; Marion, Ohio

Huber was another company with roots transcending the move from horse-powered farming to piston-powered agriculture. Edward Huber, the inventor of a horse-drawn hayrack in 1865, went on to develop a steam engine, a grain thresher, and patented dozens of other farm and industrial ideas.

This bright-minded man added to his business, with the purchase of the Van Duzen Company, a steam tractor manufacturer. Large traction engines were built under the Huber name and, in 1914, Huber introduced the company's first gasoline-powered tractor. Known as the 13-22 "Farmer's Tractor," it was a four-wheeled conventional tractor powered by a two-cylinder engine.

In 1916 a "Light Four," powered by a Waukesha engine, was introduced and in its 1921 ad Huber claimed the Light Four "As Reliable as Horses and Much Faster." It could cover an acre an hour with three plows.

Over the next 20 years, Huber manufactured a series of lightweight machines powered by Stearns, Waukesha, and Buda engines for all kinds of farming operations, but the company is best known for its early models. The company continued building tractors until World War II, when production ceased and the factory converted to the production of war materials. After the war, Huber dropped tractors from their product line but continued to build construction equipment.

Intercontinental; Grande Prairie, Texas

The Intercontinental Manufacturing Company was the builder of a single farm tractor the Intercontinental tractor Model C-26. Available in both a tricycle and four-wheeled version, it was powered by a Continental Red Seal engine and was offered with adjustable rear track, hydraulic controls, electric start, and a power take-off.

Intercontinental had manufacturing and assembly facilities in Texas and California. It

Iron Horse

This black and white ad from 1920 is for the Iron Horse Tracklaying Tractor. It offered 150 square inches of foot print on a pair of round tracked wheels. The operator in this ad looks like Luther Burbank, the world-renowned horticulturist.

attempted to export its new products and entered the Mexican market with banner headlines in upscale magazines. However, the market wasn't buying, especially when a mass of human power could be had for a few pesos a day. By the mid-1950s, Intercontinental had folded up its tractor operations.

Iron Horse; *Los Angeles, California*

The Iron Horse Riding Tractor was another small row-crop, truck farm, or orchard tractor built in California in the late-1910s. Iron Horse called it a tracklayer, but it was a wheeled tractor with a piece of track wrapped around the wheels. This design gave it maximum traction with virtually no wheel slip. The Iron Horse was one step up from a walk-behind tractor, like the Sprywheel or the Flex-Tred, but was highly maneuverable and could be used with a variety of implements.

It was used primarily in the orange orchards of southern California. However, like many of the other small California tractor companies around at the same time as the Iron Horse, it didn't last long. Interestingly, the image used in the ad appears to be Luther Burbank, one of America's greatest horticulturists, driving the Iron Horse; no reference is made in the copy.

La Crosse; *La Crosse, Wisconsin*

The La Crosse Tractor Company was the builder of the "Happy Farmer Tractor." La Crosse was formed in 1916 when Happy Farmer and the Sta-Rite Engine Company merged. The standard product for the next five years was a three-wheeled kerosene tractor built as the Model A, B, F, and G. The G was a four-wheeled version of the 12-24 Happy Farmer.

The company apparently did quite well, if you can believe this ad from 1918. It claimed to have earned over $2 million in sales in a single day at its distributors' convention. In 1921

Two Million, Two Hundred Sixty-two Thousand, Nine Hundred Seventy-five Dollars

IN ONE DAY

This was the volume of business contracted for by HAPPY FARMER Distributors at their Second Annual Convention at the Factory, at La Crosse, Wis. Deliveries will begin immediately. Manufacturing facilities are being greatly increased and a

Nation-wide Advertising Campaign

for the benefit of HAPPY FARMER Dealers will enable them to cash in on the Tractor Business in 1918. *Wire us to-day* for Territory and a sample carload of HAPPY FARMER Tractors. If your competitor beats you to it HE WINS and YOU LOSE! Remember, the Farm Tractor is the *Coming Big Business Opp rtunity.*

La Crosse Tractor Co., Manufacturers *Not Distributors* **La Crosse, Wisconsin**

Lauson

The Lauson Tractor Company emerged in the early 1910s, and as this 1918 ad notes, its 15-25 F Full Jewel Tractor was backed by 20 years of experience. The fine line ad was typical of the post-World War I tractor advertising art.

it introduced a strange tractor called the M 7-12. This unit was engineered to be driven from a drawn implement via a pair of leather straps, imitating a team of horses. In 1921 the Oshkosh Tractor Company purchased La Crosse but failed to finance the deal and the company collapsed in 1923.

Lauson; *New Holstein, Wisconsin*

John Lauson, of the Lauson Tractor Company, was the son of a German immigrant. He evolved his father's blacksmithing business, started in the 1880s, into a steam engine repair facility, adding the manufacture of traction and stationary engines for farm use.

The popularity of his "Frost King" stationary engines helped to build a cornerstone for the new tractors. Soon Lauson had a stylish 15-25 tractor available in 1915, followed by a 15-30 and 12-21.

Lauson's advertising claimed its products were "Backed by Twenty Years of Experience"; however, the combination of the Great Depression and dust bowl conditions in the American heartland brought about the collapse of the company in the mid-1930s. It was reorganized and some engine production resumed, but in the early 1940s the company was sold off to Hart-Parr which later sold it to Tecumseh in the 1950s.

Massey-Harris; *Racine, Wisconsin*

Massey-Harris, a Canadian venture, arrived on the American market in 1910 after the Massey brothers sold their interest in Sawyer-Massey. Massey then joined forces with Alanson Harris to form Massey-Harris.

The new company acquired a wide assortment of small implement companies including Deyco-Macey, an engine manufacturer from New York. They picked up the marketing rights to Wallis tractors in Canada and then purchased the J. I. Case Plow Works Company in 1928 putting them in a strong sales position.

The company continued to develop, expanding the line of Wallis tractors and developing its own line of Massey-Harris

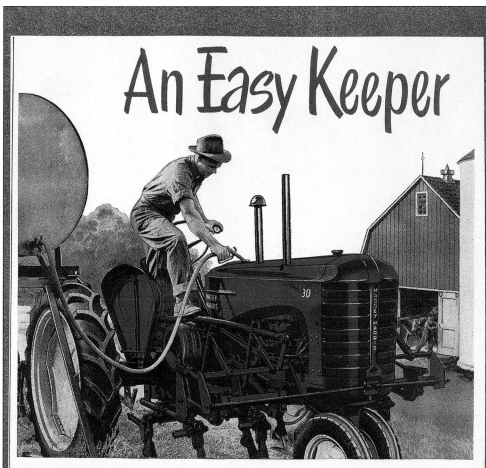

An Easy Keeper

Massey-Harris Tractors deliver more horsepower hours per gallon of fuel . . . cost less to own and operate

Actual tests prove a Massey-Harris is more economical to own and operate. Its heavy-duty high-compression engine delivers more power . . . does more work on a tank of fuel.

The Massey-Harris is designed for better farming — built to give you more years of dependable service. Its helical-gear transmission means greater efficiency . . . extra draw-bar pull for tough, hard going. Whether you're plowing, discing or cultivating, the Massey-Harris furnishes sure, steady power — geared for the job. It helps you do your field work faster, at lower cost.

It's *easier* for you to do *better* farming with a Massey-Harris, too. Shock-resistant, finger-tip steering assures you positive control . . . helps save both time and crops. Smart, streamlined design

gives you better vision . . . lets you see what you're doing without bobbing your head from side to side. Velvet-Ride Seat provides greater operating comfort . . . makes rough, bumpy fields ride road-smooth. The more flexible, more adaptable Depth-O-Matic hydraulic controls take all the work out of raising and lowering tools and implements.

There's a Massey-Harris for your farm — with a complete line of custom-built implements for plowing, bedding, planting, cultivating, mowing. See your Massey-Harris dealer the next time you're in town. Let him give you ALL the facts on Massey-Harris Tractors and Farm Equipment. You'll see why more and more farmers are swinging to Massey-Harris. For a complete folder by mail, use the coupon below.

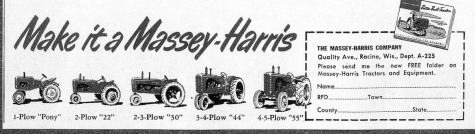

Make it a Massey-Harris

1-Plow "Pony" 2-Plow "22" 2-3-Plow "30" 3-4-Plow "44" 4-5-Plow "55"

THE MASSEY-HARRIS COMPANY
Quality Ave., Racine, Wis., Dept. A-225
Please send me the new FREE folder on Massey-Harris Tractors and Equipment.

Name..
RFD.............Town..............................
County..............................State............

Massey-Harris

The copy in this 1950 Massey Harris ad could have come from any competing maker. The company offered better fuel economy and improved seating with lower overall operating costs.

Massey-Harris

The Model 33 was typical of this sized post-World War II row-crop machine. It offered different wheel arrangements with a new 201-ci engine, a "Velvet-Ride" seat, and a "Depth-O-Matic" hydraulic system to control attachments.

tractors, including the Massey-Harris 20-30, the Challenger, the Colt, the Mustang, 20, 30, 44, and 55 models all followed in later years.

After Harry Ferguson broke away from the Fordson tractor deal in 1947, he formed a new tractor company in Detroit called Harry Ferguson, Incorporated. In 1953 he sold his company to Massey-Harris and the name became Massey-Harris-Ferguson. This purchase brought to market Ferguson's tough little TO-35 tractor, which the company soon

Owners say: "More Powerful, More Economical, Easier to Handle"

Massey-Harris 4-5 plow 55 . . . the most powerful farm tractor on wheels.

Here's power that knows its own strength — *and when to use it*

Put a Massey-Harris 55 to work on the tough jobs and watch the way it performs . . . the way it handles big equipment . . . a 4- or 5-bottom plow, a big combine, one-way disc, loaded rice cart . . . takes the fight out of land-leveling, building terraces or rice checks, handling heavy front end loaders.

It's on jobs like this that the 55 gives you a real power thrill. You sense it the minute you take hold of the wheel. You know it's there when your plow sucks to full depth and the pull stays smooth, steady . . . dependable.

Power — there's lots of it. And a host of other features you'll like too. Comfort! Vision! Elbow room and leg room! A ride that's jolt-free. Steering that takes the strain from arms and shoulders. Controls so placed that your hands and feet set themselves naturally.

Ease of handling unmatched in any other tractor!

Equally important is the amazing economy of the 55. A check of the fuel tank tells you this tractor uses fuel according to load demands . . . and gets more out of what it does use.

It's all the result of Balanced Power Design in Massey-Harris tractors . . . the perfect coordination — teamwork — between the units that *develop* the power and those that *use* it. Your dealer will tell you the whole story. See him and ask for an on-the-farm demonstration.

The facts about Balanced Power Design and what Massey-Harris owners have told us about their tractors are published in a new booklet — "Out of the Mailbag". For your free copy write, Massey-Harris, Dept. B-225, Racine, Wisconsin.

1-2 plow Pacer

3 plow 33

3-4 plow 44 Special

4-5 plow 55

MAKE IT A **Massey-Harris**
Outstanding for Economy, Power, Ease of Handling

upgraded into their own 35, 50, 65, and 85 models, all based on the original Ferguson design.

In 1958, Massey-Ferguson went corporate shopping and purchased the British diesel engine builder Perkins. This gave them a source of diesel engines and technology for their tractor line. In 1960 they acquired the Landini tractor company and its line of crawlers.

The Harris name was later dropped, but the brand name of Massey-Ferguson continued. The company has always run on the idea that variety brings business and, as the years have moved along, Massey-Ferguson has not only serviced the basic tractor needs of farmers but has specialized in advanced super-tractors, articulated models, super four-wheel drives, and specialized planting and harvest machinery.

Minneapolis; *Minneapolis, Minnesota*

The Minneapolis Steel and Machine Company appeared in the farm machinery business in 1902 evolving from numerous other small companies that had merged during the latter part of the 1880s and 1890s. It was the final blending of the Minneapolis Malleable Iron Company and the Twin City Iron Works which formed the new Minneapolis Steel and Machine Company.

Initially the company built steam engines and a mass of industrial equipment including Muenzelgas engines. Then in 1910, they expanded to sell tractors. A huge gas-powered, four-cylinder machine was built by the Joy-Wilson Company, which was promoted as the Twin City and sold by Minneapolis Steel and Machine. Some of these machines were huge with shipping weights as high as 25,000 pounds. Badge engineering was rampant in these times—companies would purchase tractors built by another company (or at least purchase the engine) and then put their own name on them. Minneapolis Steel and Machinery built tractors for Yuba, Sawyer-Massey, Reeves, Monarch, and Buffalo-Pitts, which these companies marketed as their own machines.

MS&M continued to specialize in heavyweight tractors for the next few years,

Minneapolis

The Minneapolis Steel & Machine Company claimed that its "Twin City" 12-20 Kerosene Tractor was "Dependable as the locomotive." This model was the beginning of "Twin City's" new range of lightweight tractors weighing 10,000 lbs.

FOLLOWING

Massey-Harris

The MH-50 was new for 1956 and introduced the Hydrodynamic Power systems based on the older Ferguson system. Painted in bright red, it came with the Continental Torquemaster four-cylinder engine, a Duo-Range clutch, a Double-Duty PTO, and a Hydramic Master Control.

In seconds you know...

Hydramic Power
FABULOUS MH50

No room for doubt here. You feel it instantly: the rugged power and drive of this fabulous MH50. It shows in the easy way you take off with a three-bottom plow in normal going . . . in the mountains of reserve power when you sock the shares deep in a tough spot. It's then you get *positive proof* of the dramatic difference in tractors. Quick as a wink the Hydramic-Powered MH50 takes up the fight . . . drives through with extra torque, extra traction. No engine stall, no spin-outs with Hydramic Power. Hi-torque performance — the most important of many fabulous features that are yours *only* with this exciting new tractor. Don't put off testing the MH50. Drive it . . . thrill to the miracle . . . on your farm, soon.

  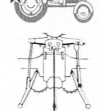  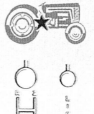 

TORQUEMASTER 4 ENGINE

Stays on the pull in the low rpm's. Gas and LP models.

DOUBLE-DUTY PTO

MH50 has both ASAE standard and speed-matched ground ratio PTO.

DRAFT MONITOR 3-POINT HITCH

Matches traction to load. Hooks up in minutes to a host of 3-point tools.

HYDRAMIC MASTER CONTROL

Raises, lowers, controls depth, holds position, regulates response.

HI-LO TRANSMISSION

Six forward . . . two reverse speeds . . matched to every job.

DUO-RANGE CLUTCH

One pedal, two stages . . . to engage engine and control live PTO.

the dramatic difference

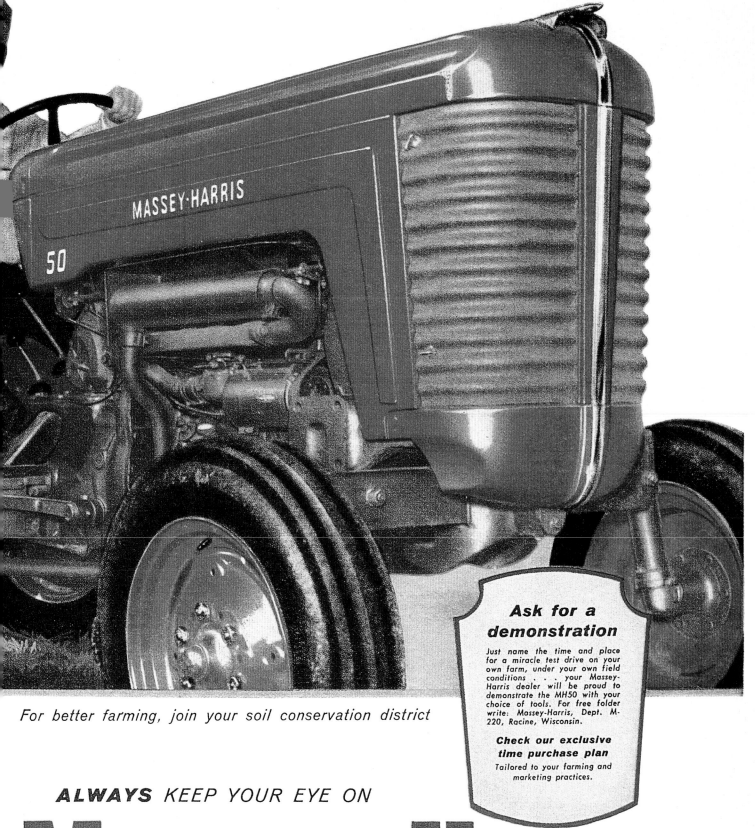

MASSEY-HARRIS

50

For better farming, join your soil conservation district

ALWAYS *KEEP YOUR EYE ON*

MASSEY-HARRIS

DIV. OF MASSEY-HARRIS-FERGUSON, INC.

Moline

The Moline Plow Company offered the Moline Universal Tractor with a price range of $765 to $1,055 in 1923, depending upon which implements were ordered by the farmer. It was one of America's first row crop tractors. It is interesting to note that in this line art ad, the tractor operator is drawn much smaller than in real life, giving the impression that the tractor is much larger.

and finally offered its first lightweight model in the mid-1910s—built by the Bull Tractor Company. Close to 5,000 were sold. A smaller "Twentieth Century" model was also built and met with similar success.

By 1920, the lightweight tractor had taken over much of the work previously done by the heavyweight models and was capable of much more. Twin City tractors were soon well known for the new 12-20 and 16-30 and its 20-25, 17-35, and 27-44 models. Interestingly, the 12-20 featured a four-cylinder engine with twin-overhead cams and sixteen valves, much like the heads used in today's high performance sports cars!

In 1920, Twin City called its 12-20 "Dependable as the locomotive" with the ad

featuring a train rushing through the background as a 12-20 neatly plowed up the ground. The Minneapolis Steel and Machine Company continued to offer a wide variety of tractors and other farm machinery through the 1920s, and at the end of the decade, with two other companies, merged its money, facilities and patents to form the Minneapolis-Moline Power Implement Company.

Moline; Moline, Illinois

The history of the Moline Plow Company is a mass of complex buyouts and the merging of business interests that go back to 1852 when

Henry Candee and Robert Swan formed Candee and Smith.

With six new partners the company reformed in 1870 as the Moline Plow Company. In the ensuing 20 years it purchased a steel company, the Mandt Wagon Company, the Henney Buggy Company, and the Freeport Carriage Company. Not satisfied, Moline barreled ahead buying up the Monitor Drill Company, McDonald Brothers Scale Company and the harvester manufacturer, Adriance, Platt and Company.

In the mid-1920s Moline experimented with a series of tractors, but not getting the results required, they simply purchased the Universal Tractor Company. This unusual machine looked like a giant garden cultivator with a pair of huge drive wheels up front, no bodywork, and seating position for the driver at the rear over the implement. The Moline-Universal Plow looked like the 1917 Allen Water Ballast tractor which used water-filled front wheels to give added weight to improve traction.

Moline upgraded the design with a four-cylinder engine of its own and a line of attachments including plows, discs, corn harvesters, and power take-off capabilities. The Model D was the company's most successful version of the Moline Universal Tractor and was sold from 1918 to 1923 with 17 horsepower at the drawbar.

Interestingly, the 1919 advertising depicted the tractor much larger than it was; the artist drew in the operator at half-scale so that the machine looked gigantic! The later 1923 ads had the scale about right and, with a selling price of $765 and with a two-row cultivator, it was an attractive deal.

During this time the Willys Corporation purchased a huge slice of the company shares, while Moline purchased Independent Harvester Company and then merged with Root & Vandervoot Engineering in 1921.

Slow sales, caused by the huge postwar depression, eventually forced

Monarch

The Monarch Tractor Company built crawler tractors until 1928 when Allis-Chalmers purchased the company. This 1919 ad for the 18-30 "Neverslip" crawler featured a rear-positioned driver and seems to completely ignore the operator's safety. The model was never popular—could it be because the driver's vision was impaired by the positioning of the engine and the fuel tank?

Made Possible the Nelson Four-Wheel Drive

NELSON TRACTOR, SHOWING BELT PULLEY ATTACHED FOR USE AS A POWER PLANT

THE NELSON is not an adaptation of any other tractor. It was designed by the Nelson Engineers to meet the known requirements of the Farmer or commercial user, and was developed carefully and scientifically unit by unit.

All material used in its construction is the best to be had for the work it is intended to do, and every part is designed to give years of satisfactory service.

That simplicity without loss of efficiency, and power without bulk, are the keynote of its great success.

PULLEY power with the Nelson Four-Wheel-Drive Tractor is secured direct from the main transmission shaft, which is connected directly to the engine by the easily operated main clutch that drives the tractor. The pulley is placed snugly up against an oversize bearing, and all strain and friction are reduced to the minimum, making the pulley power equal to the maximum power of the engine. In addition, the tractor can be driven backward and forward while the pulley is turning and the belt is still in place, in order to bring it into perfect alignment and the proper tension.

WE recommend the Nelson Tractor to the farmer, because it will work for him steadily all the year round regardless of soil conditions.

It has been tested in the mountains and in the swamps, and has worked steadily in sand and snow, under the worst possible conditions of heat and cold. It has taken the place of mules on the rice fields, and has displaced in many instances the horse in the logging camps. Contractors and road builders are finding it all that we claim for it, and are constantly adding Nelson Tractors to their equipment.

Manufacturing Facilities Guarantee Prompt Deliveries

Nelson

In 1919, the Nelson Corporation from Boston offered the Nelson Four-Wheel Drive Tractor. The Nelson delivered great traction and ground clearance but apparently had problems with its steel wheels on this version. The tractor was extensively tested in mud, snow, and sand.

Moline Plow into a three-way deal with Minneapolis Steel and Machinery and the Minneapolis Threshing Machine Company; the resulting company was Minneapolis-Moline.

Monarch; *Watertown, Wisconsin*

The Monarch Tractor Company was started by tractor engineer and designer, B. Smith, in 1913. From a small Lightfoot 6-10

THE·REEVES·CROSS COMPOUND DOUBLE CYLINDER PLOWING ENGINE

It is a big engine: big every way: big in power: big in ability to do work: big in size of all material parts: big cylinders: big boiler: big wheels: big gearing—made for strength: made to last. It's the one you want for any purpose for which a traction engine can be used. More Reeves Cross Compound Double Cylinder Engines are used successfully for plowing than all others combined. It is equally as popular for threshing, hulling, shelling, saw milling and other work requiring power. We want to tell you all about it. If you want to know, write for catalog. It's free.

REEVES & COMPANY · COLUMBUS · INDIANA

Reeves
Reeves & Company began building steam traction engines in 1885 and became famous for the cross compound double- cylinder plowing engine featured in this 1909 company advertisement. The Reeves machines were huge; the company was eventually sold to Emerson-Brantingham in 1912.

crawler powered by a Kermath four-cylinder engine, Smith developed six models which were built in five different company-owned facilities. The company reorganized in 1919, changing its name to Monarch Tractors Incorporated, and opened a new factory in Brantford, Ontario. The company was again re-formed in 1924 as the Monarch Tractor Corporation.

From the first small Lightfoot model emerged a larger Erd four-cylinder-powered "Neverslip" 12-20, followed by the 18-30 (as seen in the ad), and then the short-lived 30-55. Moving on to bigger and better models in 1924, Monarch

introduced the 60 horsepower Beaver-engined Model D with its enclosed cab and then the 75 model as its largest crawler. Powered by a 75 horsepower four-cylinder LeRoi engine it looked very much like a Caterpillar of that vintage.

Buy-outs were common in the farm equipment business at the end of the 1920s, and in 1928 Monarch was purchased lock, stock, and barrel by Allis-Chalmers.

Nelson; *Boston, Massachusetts*

Since 1900, the idea of a four-wheel drive tractor had been toyed with by a number of tractor

manufacturers. The Nelson Blower and Furnace Company designed its first and only tractor as a high-ground-clearance, four-wheel drive model fitted with huge self-cleaning steel-spoked wheels. The tractor was designed to be used as a power unit for farm work and as a stationary engine. Powered by a Wisconsin four-cylinder engine, the Nelson offered 15 horsepower at the drawbar and 24 brake horsepower at the belt.

In its advertising Nelson claimed they had tested the tractor "in the mountains, swamps, sand and snow and under the worst possible conditions of heat and cold." It sounded like a great machine, but like many other tractor manufacturers of the time, the company built other products and due to the oversupply of tractors in the 1920s, Nelson closed up its tractor division after only a couple of years.

Reeves; *Columbus, Indiana*

Reeves & Company was one of the nation's earliest steam engine builders, opening shop in 1874. At first it built stationary powerplants for farms and industry. Its traction engines emerged in 1885 equipped with the Clay valve, a device that offered a more economic utilization of steam which, in turn, enhanced engine performance. Advertised as the Reeves-Cross Compound Double Cylinder Plowing Engine, it offered everything "Big" as its sales advantage. It was apparently the all-in-one tractor, good for "plowing, threshing, hulling, shelling, saw milling and any other work requiring power."

In 1910, Reeves offered a new kerosene-powered four-cylinder tractor that was built for them by the Minneapolis Steel and Machinery Company, the builders of Twin City tractors. The Reeves unit was almost identical to the Twin City tractor, and when Emerson-Brantingham took over Reeves, this tractor continued to be offered for a number of years.

Rock Island; *Rock Island, Illinois*

The Rock Island Plow Company was a simple farm equipment business from its beginning in 1855 when the founders, Bulford and Tate, started manufacturing farm equipment. The company flourished into the twentieth century, selling an ever-increasing line-up of plows, harrows, cutters, harvesters, and cultivators.

In 1914, Rock Island took over the distribution of the Heider Manufacturing Company's farm tractor line and in 1916, purchased the company outright. The Heider tractors had become so popular that the supply demand could not be met, so Rock Island did the next best thing—built the tractors itself.

The Heider tractor featured the engine mounted over the rear axle and just ahead of the driver. The drive system used a roller-chain drive

Rock Island

The Rock Island Plow Company was an implement manufacturing company until it took over the distribution of Heider tractors in 1914. Within two years it had bought out the Heider Company so that it could sell a tractor to pull its plows. This 1919 ad shows the 12-20 Heider and the Model D 9-16.

Russell

Russell & Company began building tractors in 1909 but closed its facility in 1927. In the intervening years, Russell built a broad range of steam- and gasoline-powered tractors including the tractor pictured here. The "Little Boss" had a four-cylinder Waukesha engine and was rated as a 15-30 horsepower model.

which drove the rear wheels from a primary cog just ahead of the rear wheels. By 1920 the Heider models had seven-speeds forward and reverse using Heider's Patented Friction Drive.

Over the next 15 years, Rock Island developed a successful and ongoing line of lightweight tractors powered by Buda, LeRoi, and Waukesha engines.

In 1926 Rock Island's advertising claimed that its 9-16, 2-20, and 15-27 tractors "run as smoothly as an electric fan." In 1929, when the Model D 9-16 was eliminated, the Heider brand name ceased to appear on Rock Island products. Rock Island managed to survive the Great Depression of the early 1930s and the dust bowl years, only to be bought out by J. I. Case in 1937.

Russell; Massillon, Ohio

Russell & Company, like many other farm equipment manufacturers, had started off in another business at the turn of the century. In this case they were furniture manufacturers. Their first venture into steel was with grain threshers, but these were primarily of wooden construction. However, in 1885, Russell started building its own steam engines which featured "shifting eccentric valves."

Using this steam engine, they built a large traction engine that apparently met the mark rather well as they stayed in production right into the 1920s.

Russell entered the gas tractor business in 1909 with an American version of a large British tractor. Russell claimed that its new "American Tractor" delivered 44 horsepower, weighed 17,500 pounds and sold for only $2,400. It continued with this size of steam traction engine, refining it further, and renamed it the "Giant."

By 1914 Russell had entered the light-weight tractor business with a three model

SAMSON SIEVE-GRIP TRACTOR

(4 Cylinder Engine)

(Registered and Patented in U. S. and Foreign Countries)

FOR ORCHARD, VINEYARD AND GENERAL FARM WORK

Designed for an 8-Horse Drawbar Pull With Surplus Power to Pull Its Load Up Any Reasonable Grade or in a Pinch. 25-30 h. p. for Any Stationary Work.

GENERAL SPECIFICATIONS

Sieve-Grip Wheel with Road Band in Place.

Sieve-Grip Wheel Showing Grips and Cutting Rings.

TRACTOR.
LENGTH OF TRACTOR—12 feet 5 inches.
WIDTH OF TRACTOR—60 inches.
HIGHEST PART ABOVE GROUND—50 inches.
TRACTION WHEELS—Diameter, 48 inches.
FRONT WHEEL—Diameter, 32 inches; 12 inches wide. Has 3 guide rings.
FRAME—I-beams and crucible steel.
TRANSMISSION—Vanadium steel gears, enclosed and running in oil.
DRIVE—Floating axle with hub drive to crucible steel traction wheels.
SPEEDS—Two ahead and two reverse, 2 miles and 4 miles per hour. Can be varied with speed of engine.
CLUTCH—Samson improved marine; double expansion; runs in oil.
LEVERS—Two, one for gear shift, one for go ahead and reverse.
BRAKES—Independent foot brake for each traction wheel for making short turns.
PULLEY—20-inch diameter, 8-inch crown face, 1⅜-inch bore. Speed, 250 to 300 r. p. m.
WEIGHT OF TRACTOR—7,500 pounds.

POWER PLANT.
ENGINE—Samson four-cylinder L head Tractor Type, four cycle.
CYLINDERS—5-inch bore by 7-inch stroke.
CRANK SHAFT—Diameter, 2½ inches, 1-inch offset.
CRANK PIN—Diameter, 2½ inches by 2¾ inches.
FIVE CRANK SHAFT BEARINGS—Total combined width, 17 inches.
PISTON PIN—1 15-16-inch diameter and hollow. Three oiling methods to piston pin.
VALVES—Inlet and exhaust are mechanically operated. (Valves are lifted plumb, without side strain, and are enclosed.)
VALVE CHAMBERS—Water jacketed and provided with caps for easy access to valves.
CLEANING PLUGS—In center of cylinder heads, easily removed.
ENGINE SPEED—525 to 575 r. p. m.
BRAKE HORSE POWER—25 to 30.
IGNITION—High-tension magneto mounted on engine.
CARBURETER—Has no moving parts.
PUMP—Rotary. (Bronze.)
RADIATOR—Tubular, with fan.
LUBRICATION—Mechanically operated forced feed oiler, with tubes leading to each part to be lubricated.
FUEL—Engine distillate. Fitted to use kerosene, when ordered.

ENGINE AND TRANSMISSION COMPLETELY PROTECTED FROM DUST OR DIRT.

WHEELS DO NOT SLIP IN WET OR SANDY SOILS

SAMSON NON-KICK ENGINE SAFETY STARTER.   YOU CANNOT GET HURT

YOU WILL BE GLAD TO BE A SAMSON OWNER.   Ask for Catalogue P. A.

WE MANUFACTURE

SAMSON Horizontal Stationary Oil Engines, 2 to 40 h. p.

SAMSON Vertical Oil Engines, 16 to 200 h. p.

SAMSON Marine Oil Engines, 1, 2, 3 and 4 Cylinder, 4 to 200 h. p.

SAMSON Centrifugal Pumps, 25 to 50,000 Gallons per minute capacity.

SAMSON Pumping Plants, Complete, installed and guaranteed.

SAMSON IRON WORKS

Cable Address
"Samson-Stockton"

STOCKTON, CAL., U. S. A.

Samson

The Samson Sieve-Grip Tractor featured unique steel wheels which Samson claimed "Do not Slip in Wet or Sandy Soils." This 1914 ad shows the 8-25 horsepower model designed for orchard, vineyard, and general farm work. At this time the Samson Iron Works was based in Stockton, California.

line-up. These new tractors featured conventional tractor styling with enclosed engines and roofed cabs and were known as the Russell Junior, The Little Boss, and The Big Boss. These tractors offered more power for more money. In 1920 Russell's advertising claimed that buyers should "Demand Reliable Farm Power."

Russell was another of the tractor companies which the 1920s postwar Depression sent into bankruptcy. It closed up shop in 1927, although spare parts continued to be available for another 15 years through the Russell Service Company.

Samson; Stockton, California

The Samson Iron Works in Stockton, California, was one of the mass of tractor and farm equipment companies in the Stockton area at the beginning of this century.

Samson was a manufacturer of stationary and marine engines, pumps, and pumping plants and entered the tractor business with their Samson Sieve-Grip Tractor in 1914. The new tractor offered 6 drawbar horsepower with 12 horsepower for stationary work. The tractor also featured an unusual low goose-neck frame, tricycle wheels, and a single-cylinder engine.

A second model with an enclosed engine powered by a four-cylinder engine was offered with 8 horsepower at the drawbar and up to 30 for stationary work.

The logo claimed "The Strength of Samson is in Every Part" with an image of Samson in the middle. However, the Samson company's great claim to fame was the Sieve-Grip wheels which "do not slip in wet or sandy soils." General Motors Corporation also thought the idea had merit; it bought out the company. In 1919 Samson introduced a new model called the "Iron Horse" a row-crop tractor that featured a chain and sprocket drive and a track width wide enough to span two standard crop rows.

GMC had purchased the Janesville Machine Company of Janesville, Wisconsin, and moved the Samson Company to the same location in 1920. Almost immediately they offered a new GMC-designed Model M farm tractor which looked surprisingly like Henry

SAWYER MASSEY

PISTON PINS. The piston pins are of generous size, being $1\frac{5}{8}$ inches in diameter, made of steel, case hardened and ground. They are provided with bronze bushings in the piston, which can be renewed when required.

CLUTCH. The clutch is of the expanding shoe type, is self-locking, and is provided with friction shoes of hard maple which can be easily replaced. For adjustment, a turn-buckle and locknut are used, which enables taking up the wear on the shoes caused through friction. This clutch is used for forward and reverse work, as well as for belt work.

BEVEL GEAR CASE. This is cast in two pieces and has a hand hole for inspection purposes covered with a plate, which can be quickly removed. The bevels are free on the pulley shaft. The pinion runs between the two bevels, and to reverse, a dog clutch is used which operates between the two bevels into one or the other gears. The shafts of the gear case are provided with four double row annular ball bearings, which take both radial and thrust loads. These bearings prevent any side motion, and are very much superior to the babbitted bearings, as they insure that the bevel gears will always remain correctly aligned. The bevel gears are made of steel with machine cut teeth. The case is dustproof, and the gears run in a bath of oil which insures minimum wear on the parts and helps to transmit the power with the least possible friction loss.

SPEED OF TRACTOR. The Sawyer-Massey Gas Tractor has two speeds, one of 2 miles per hour, the other of $3\frac{1}{2}$ miles when the motor is running 600 revolutions per minute. Speed is changed by means of an interlocking lever from the operator's platform, so that it is impossible to have both gears in mesh at the same time. Speed can also be increased or decreased at the will of the operator by increasing or decreasing speed of motor.

GEARS. The train gears are placed inside the frame and have 4-inch face and $1\frac{3}{4}$-inch pitch. The low speed pinion is cast steel. The intermediate gear has a bronze bushing 10 inches long.

The traction wheel gears are 5-inch face and $2\frac{1}{2}$-inch pitch. All pinions are of cast steel. The bull pinions are supported by a frame bearing close up to the gear, which leaves no overhang as in other makes, and these gears are perfectly aligned and smooth in their running, which insures almost noiseless operation.

73% of weight carried
on Rear Wheels

Sawyer-Massey Gas Tractor—Left View
25-45 Horse-Power

Stinson

Stinson Tractor began production in 1917 and lasted until 1922; the company built only three models. This 1918 ad featured the 18-36 model powered by a Beaver four-cylinder engine. Its offset tricycle chassis was designed for row crop farming with a single wheel up front which apparently offered an exceptional turning circle.

Ford's Fordson. However, the economic downturn in the early 1920s forced the closure of the Samson Tractor Company.

Sawyer-Massey; Hamilton, Ontario, Canada

The story of the Sawyer-Massey company matches the history of many of the American tractor companies that emerged from the middle of the last century. Its roots were in Hamilton, Ontario, in a blacksmith's shop. The company's subsequent interest in horse-powered farm machines and steam followed along like that of the Moline Plow and the Rock Island Plow Company.

Sawyer entered the stationary steam engine market in the 1860s, continuing to expand its product line and improve upon its steam engine designs. In 1890 the Massey brothers entered the picture when they purchased a massive slice of company shares which resulted in the formation of the Sawyer-Massey company.

By the turn of the century the company was also building gas-powered models, and the 1907 model featured in the ad is the 25-45 Gas Tractor powered by a four-cylinder engine and carrying 73 percent of its weight over the rear wheels. The company continued on its course with both steam and gas, building steam engines with outputs close to 75 horsepower.

Eventually the Massey brothers pulled out of Sawyer-Massey to concentrate on gas-power machines and the new Massey-Harris organization. The Massey name remained and the company went on to specialize in heavy-duty construction equipment, eventually folding in the 1930s.

Stinson; Minneapolis, Minnesota

Two brothers, Charles and Leslie Stinson, teamed up with Richard Ruemelin to form the Stinson Tractor Company. Ruemelin resigned his engineering job with Minneapolis Steel & Machinery to serve as the secretary of Stinson Tractor in 1916.

One of the first prototypes to be tested in the field was a tractor with a movable steering position. On the left side it allowed a short

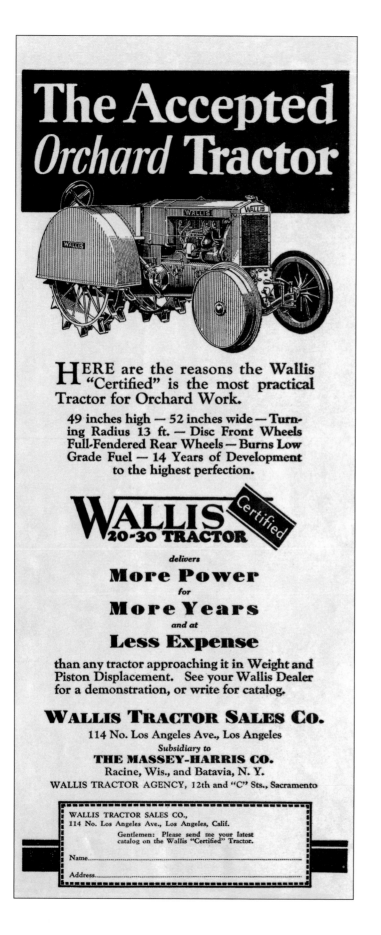

Wallis

In 1928, Wallis was purchased by Massey-Harris, although specialty tractors were still offered. According to the company, these tractors were "Certified" as the most practical tractor for orchard work. The orchard version of the Wallis 20-30 was only 49 inches high and came with disc front wheels and fully fendered rear wheels to protect trees and bushes in the orchard.

turn, center steering was used for road work, and on the right side it made plowing easier.

The improved Stinson Model 18-36 was built by Imperial Machinery Company of Minneapolis. It featured a "larger four-cylinder engine with bore and stroke of 4 3/4x6 inch with all gears enclosed in dust proof housing and operating in a bath of oil. A one piece semi-steel frame with main bearings machined into the frame assured proper alignment of the bearings for the life of the tractor."

Over the years Stinson had contracts with two different companies: the Gile Engine Company, which built the Stinson tractor in 1917; then the Imperial Machinery Company in 1918. The following year, Stinson moved production to Superior, Wisconsin, but there the company only survived until 1922.

Wallis; *Racine, Wisconsin*

Started in 1902, Wallis Tractor was an early player in the tractor business, creating interesting new ideas, including the unit-type design—a lightweight tractor without a chassis. This frameless idea is still in use today. By 1913 it had several models on the market including the Cub and Bear models. H. M. Wallis was president of J. I. Case Plow Works Company, but the Wallis Tractor Company was run as a separate enterprise. The 1916 Cub was the first frameless tractor with a "Unit-Frame." The one-piece frame had the crankcase and transmission housing built from a boiler plate, and this formed the backbone of the tractor. It rolled on a tricycle layout and was powered by a vertical four-cylinder engine.

The idea evolved further and by the time J. I. Case Plow Works Company took over Wallis, it had adapted the design of a one-piece crankcase and final drive.

Waterloo

This ad for the Waterloo Boy distributor was placed by W. L. Cleveland for the opening day field demonstration of the Waterloo Boy tractors at the Davis, California experimental farm in 1919. The Waterloo Boy was called the "One-man Tractor," and this model N was one of nearly 20,000 that the company manufactured between 1917 and 1924. Powered by kerosene, its two-cylinder engine developed 12 horsepower at the drawbar and 25 at the belt.

In 1920, Wallis shipped an O model to England for field testing at the Royal Agricultural Society's annual show. It won an outstanding gold and first prize after seven days of testing in three different plow classes.

Like almost all the tractor advertising in the 1920s, Wallis praised its tractors for efficiency and economy with headlines like "Farm More Acres Per Hour."

Wallis used J. I. Case as its distribution arm and continued developing new and improved tractors, including the "OK" which was introduced in 1922. The OK was an immediate hit with its three-plow rating for a two-plow weight.

In 1928, the company was sold in an involved deal with Massey-Harris, J. I. Case Plow Works, and the J. I. Case Threshing Machine Works. Eventually the deal was resolved with Massey-Harris owning Wallis and the Case Plow Works, but selling the Case name back to Jerome Case.

Waterloo; *Waterloo, Iowa*

The Waterloo Gasoline Engine Company was started by one of the most inventive of tractor engine builders, John Froehlich. His ideas and improvement on the Van Duzen gas engine were notable and patented. He left the company when it decided to pursue the stationary engine business rather than the tractor business. Waterloo continued with its own engine designs by Louis Witry and later Harry Leavitt.

Eventually, the company entered the tractor business in 1912, featuring the "One Man Tractor." Built solidly with a weight of around 9,000 pounds, the Waterloo Boy used a cross-mounted engine.

Waterloo Boy entered the lightweight tractor business in 1914 with its 12-24 model. Its instantaneous success was noted by others and emulated across the market place; the 2-24 remained in production for ten years. The more powerful "R" and "N," models were introduced in 1916 and looked much like the original 12-24. Both four- and three-wheeled versions were offered including some special models for orchard work with lower chassis heights. The most successful of these was the model "N," which stayed in production until 1924 with 20,000 units. The quality and reliability of Waterloo Boy tractors made them the industry standard by which others were judged.

John Deere took note of this and purchased the company in March 1918. This purchase turned John Deere overnight from a farm implement manufacturer to a full-line farm equipment builder and gave the company a jump start in technology to develop a tractor line of its own. However, Deere continued to build and sell Waterloo Boy tractors as a separate line until the mid-1920s.

Willys-Overland; *Toledo, Ohio*

The Willys-Overland Corporation could see the end of World War II approaching and with the massive production lines it had available to build its amazing little Jeep MB 4X4, it looked to agriculture to take some of those vehicles and use them on the farm.

Designed by Karl Probst and Harold Crist, and production engineered by Barney Roos at Willys, the Jeep had been one of the key elements in the Allied victory in World War II. Both Ford and Willys built the Jeep during the war, but with the cessation of hostilities, Ford immediately abandoned its Jeep GPW production line and returned to the production of civilian automobiles.

It was a different matter at Willys. It had already civilianized the Jeep MB, hoping to sell it to returning troops who knew the Jeep's

The Sun Never Sets On the Mighty "Jeep"

JEEPOTENTIALITIES

Tough, irrepressible, battle-stained—the mighty "Jeep" is today doing its appointed task on the battle fields of the world, *as planned.*

That this mighty "Jeep" is speedy enough, powerful enough, versatile enough and rugged enough to meet the stiffest demands of war is not an accident.

The "Jeep" was designed, powered and strategically engineered expressly as a *war* machine—to perform and *keep going* no matter *when, where* or *what* the demand—and to stand up.

The same skilled craftsmen who were able to produce a "Jeep" equal to the harsh and unremitting demands of a world-wide war, have developed plans to adapt it to greatly expanded usefulness on farms, in industry, in small businesses and among individuals with many widely diversified needs. Have you considered Jeepotentialities—after the war?

★　★　★

We have received many letters asking us about buying "Jeeps" after the war. Among the numerous "Jeep" uses suggested in these postwar plans are, operating farm implements, towing, trucking, road work, operating power devices and many personal business and pleasure uses. Does the "Jeep" figure in *your* postwar plans? Willys-Overland Motors, Inc., Toledo 1, Ohio.

Willys *Builds the Mighty* 'Jeep'

Willys-Overland

While some may not consider the Willys Jeep a tractor, the company certainly thought the little 4x4 could do the job in the immediate post-World War II years. Spearheaded by Barney Roos, Willys had to do something with its excess capacity to build Jeeps at the close of the war. This Willys Jeep ad from 1945 planned for the postwar years when the little fighting machine would become a working man's miracle, plowing fields and running about the farm. This powerful image reflected the dream of turning guns to plowshares.

abilities. It seemed like a good plan. In 1944–45 Willys ads stated, "The Sun Never Sets on the Mighty Jeep" and sales were focused on farmers, informing them that this small, all-purpose vehicle could power their farm implements, run to town, and haul small loads.

During this time the American Bantam Company took both Willys and Minneapolis-Moline Power Implement Company to court for the ownership of the Jeep name. However, American Bantam won a somewhat hollow victory as after several years of investigation and a second trade name court battle, the Federal Trade Commission eventually assigned American Bantam Company the title of originator and developer of the Jeep but gave Willys the right to use the name and to build the vehicle. Minneapolis-Moline Power Implement Company had used the Jeep name on one of its earlier products and hoped to keep it for future use.

Barney Roos envisaged a massive market for an agricultural Jeep and commissioned 22 prototypes, code-named CJ-1A. The CJ designation was for "Civilian Jeep" and the production models that emerged from Willys in August 1945 were MB versions that included a tailgate, a side-mounted spare, larger headlamps, and an external fuel cap. It was introduced as the CJ-2A and sold for $1,090.

In its 1947 ad, Willys called the Jeep "The Work-Horse of the World" showing it hauling farm products to market, pulling a disc plow, being used as a mobile power unit and driving through mud.

It was a well thought out idea, but the Jeep never made the grade as an all-purpose farm tractor/truck. Authentic tractors were needed for the rigors of farm work. However, the Jeep met with success as an all-purpose recreation and farm vehicle and is still considered untouchable in that arena today.

Yuba; Marysville, California

The Yuba Manufacturing Company was another of the small tractor builders centered in the Central Valley of California. Yuba commenced as Yuba Construction and entered the tractor business with the purchase of the Ball Tread Company, a crawler tractor manufacturer in Detroit, Michigan.

Ball Tread tractors featured a set of large steel balls on which the tracks rolled, thus the name. When Yuba took over Ball Tread it had two models in production, the 12-25 and the 8-35. Both of these models continued as the line expanded with the 12-20, the 20-35, and the 40-70 models. Many road construction companies liked the Yuba and they were used extensively for agricultural work in the development of the "Salad Bowl" in central California.

Yuba crawlers weighed in the 5,000- to 6,000-pound range and sold for as much as

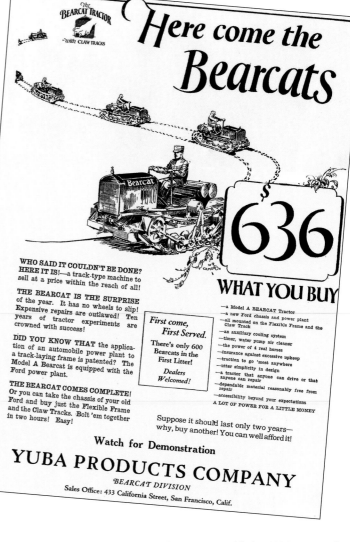

Yuba

"Here come the Bearcats" bannered the Yuba Products Company ad for its 1923 Bearcat crawler conversion. Based on a Model A Ford, the conversion offered a lightweight crawler tractor for only $636 and noted that, "There's only 600 Bearcats in the First Litter!" This small crawler tractor featured a claw track, new Ford chassis, and a powerplant that was of "Utter simplicity in design."

$6,250. All these Yubas were set with tricycle wheels and track positioning. In 1923 the small Bearcat crawlers were introduced with "Claw Tracks" which functioned like a Caterpillar crawler. About 600 of these were built the first year and Yuba managed to stay afloat into the early 1930s, but was finally forced to close its doors in 1931 under the burden of the Great Depression.

Index